WINDOW GLASS DESIGN GUIDE

edited by Denis Philip Turner MISTC

The Architectural Press Ltd: London

Nichols Publishing Company: New York

Acknowledgements

This book was compiled from the work of members of the Pilkington Technical Advisory Service, Merseyside. Chapters 1 and 2 first appeared in *The Architects' Journal* from 10 December 1975 to 28 July 1976; Chapter 6 first appeared in *Light and Lighting* and *Environmental Design* in January-February 1975; Chapter 7 was first published in *The Architect* in May 1976 under the title 'Glazing and heat gain in offices' and is reproduced here by kind permission of the editor.

First published in book form by
The Architectural Press Ltd: London 1977
ISBN: 0 85139 709 3

Library of Congress Cataloging in Publication Data

Main entry under title:

Window glass design guide.

Bibliography: p109
Includes index.
1. Windows — Handbooks, manuals, etc. 2. Glass — Handbooks, manuals, etc. I. Turner, Denis Philip.
TH2275. W56 7217.8 77-6739
ISBN 0-89397-028-X

Printed in Great Britain by Diemer & Reynolds Ltd, Bedford.

Contents

Introduction

In recent years a greatly improved performance has been demanded of the building fabric in general and of the window in particular. The need for improvement has been emphasised by sudden, large increases in the costs of fuels and by realisation of the need to conserve our energy resources. Advances in glass technology have made it possible to manufacture glass products that suit these new design requirements, but visual, environmental, and structural performances of the newer window glasses are necessarily complicated and the design techniques needed to make best use of the products are not widely known.

Many of the items of information relevant to modern window design have been published, but published separately and, in 1975, the time seemed to be appropriate for collecting them in the Window Glass Design Guide that was published in *The Architects' Journal* in various issues from 10 December 1975 to 28 July 1976. The Guide considered the principal environmental factors that affect the selection of the glass for any particular situation and itemised the commercially available window glasses and their performances. The products and components used in building cannot be considered and selected in an isolated design process and, because glass is no exception, reference was made to *The AJ Handbook of Building Enclosure* for information about the external envelope, and to *The AJ Handbook of Building Environment* for the internal environment.

In the limited space that was available for the original publication of the Guide, several of the prediction techniques could be presented in outline only and other important sources had no more than a brief reference. This new book corrects those shortcomings by bringing into one volume the original design guide, the worked example that completed it, the associated design techniques and other recently produced items that are relevant to the design of windows.

The book was compiled from the work of members of the Pilkington Technical Advisory Service, a service that provides guidance on the properties and applications of glass. As far as possible where products are discussed, the information given is of a general, world-wide nature and not restricted to the products of any particular company. There are, inevitably, isolated places, usually where no national or international standard or accepted practice exists, where the authors have had to rely on their experience of Pilkington methods or products. All designers should be aware that glass manufacturers provide a technical advisory service on the properties and performances of their own products and advise on design problems, product selection, and methods of installation

In scope, this book confines itself to the discussion of flat glass for windows in the building facade and of the interactions between the performance of the glass and the properties of the window frame and the building. It does not deal with moulded glasses, such as channel glazing or glass blocks, nor with door and internal glazing and the attendant concern for security, fire safety or the risk of personal injury. It is not a 'design guide' in the sense in which *The Architects' Journal* uses that term to describe a highly organised checklist that makes reference to technical studies and information sheets. Here, the technical data are often incorporated in the design guide itself.

It is intended to deal mainly with the *internal* functions of the window: that is to say, the ability to provide the interior spaces of the building with light, sun, air and view according to the desires of the occupants and to shield them from such unwanted aspects of the external environment as dust, noise, rain or excessive heat or cold.

But windows have an equally important set of *external* functions. They can in suitable instances provide passers-by with attractive views into the building (eg into commercial buildings; but possibly also into houses, though this is considered less desirable by the English householder than by, for instance, the Dutch); and they make a powerful contribution to the external appearance and character of building facades **1** to **3**.

While the latter aspect cannot be examined in detail here, it is necessary to say something about the potential conflict that may develop between the internal requirements of the window, which are largely functional, and the external, which are largely to do with appearance and character).

When the two sets of requirements pull in opposite directions, the architect can land himself in problems that are almost insoluble—his entrancingly transparent glass pavilion becomes a heat trap in summer and an ice-box in winter; his elegantly simple large pivot windows are found, by occupants, to give a simple choice between either too much ventilation, or none at all. *How is the designer to avoid choosing fenestration patterns which throw up this sort of conflict?*

In the past, AJ design guides have taken the line that external character should grow rationally out of a solution to internal problems—the architect should begin by analysing carefully each internal space in the building in terms of light, air, sun, view, privacy, and acoustic requirements, decide on appropriate window arrangements that might suit these internal requirements, and only then allow images of external appearance to invade his mind.

Today we must concede that very few architects are really able or willing to work in this way. Whether we like it or not, designers start with images of external appearance, of the visual contribution made by the window to the facade and the facade to the streetscape, and there is a lot to be said for such an approach. Streetscape and the visual and spatial quality of spaces and facades are of paramount importance to most people as their love of medieval townscapes demonstrates more forcibly than words could do.

A more sensible approach might be to accept that architects begin to think about external appearance at a very early stage in design, to steer them towards sensible concepts and away from silly ones and to help them deal effectively with such problems as might remain even after a fundamentally sensible fenestration pattern has been selected.

The forthcoming *AJ handbook of environmental design* will attempt to come to grips with this problem across the whole range of environmental performance (heating, lighting, acoustics, ventilation). Meanwhile, this more limited design guide can give only the following advice on dealing with the external functions of windows in such a way that intense conflicts with internal functions are avoided:

● The larger the window, the more acute the internal problems of thermal control, noise control, and economy of operation are likely to be. From the strictly functional point of view, glazed area should seldom exceed 20 to 40 per cent of total facade area, **3** (though there will of course be exceptions, where other considerations take precedence).

● If windows have to be large, effective shading devices can do much to ameliorate excessive sky-glare and heat build-up, but they will of course not prevent heat-loss in winter, **4**, **5**.

● Beyond reducing window size and fitting shading devices, the next most effective option is the use of special glass, **6**. Leaving aside other reasons for specifying special glasses (eg increased privacy or mirror-like appearance), they will allow windows to be rather larger than might otherwise be the case from the thermal control and glare-avoidance point of view.

Within this overall framework of a sensible selection of basic building concepts, so that insoluble problems at detail design stage are avoided, the design guide is intended to provide the architect with a feel of window glass: its nature; its performance; the range of prediction techniques that may be applied; and the implications of window design and glass selection in terms of building structure, building services and the comfort of the occupant.

1

2

3

4

5

1 *Curved facade off Sloane Square, London, as part of streetscape.*
2 *A glass wall for Olsen by Foster Associates.*
3 *A building with a definite fenestration pattern by Faulkner Brown.*
4 *GLC island block, County Hall. External view of blinds.*
5 *GLC island block. Internal view when blinds drawn.*

6 *Special glass used to provide an environmental package, including heat and light control, together with a weather seal.*

What is meant by 'window glass'?

Window glass may be considered as a vitreous silicate made from silicon dioxide, sodium oxide, calcium oxide and sodium carbonate. In manufacture when the batch mixture is heated the sodium carbonate melts and reacts with the sand, forming sodium silicates. Limestone and dolomite are added to form a soda-lime magnesium silicate, for durability against rainwater and acids. Their addition complicates manufacture and is kept to less than 12 per cent.

The other constituents are approximately silica 73 per cent, alumina 2 per cent and soda 13 per cent (other ingredients are added when special properties are called for). This mixture makes a typical clear glass.

7 *The float process.*

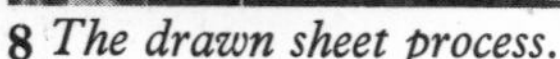

8 *The drawn sheet process.*

9 *The rolled process.*

Heat absorbing qualities are accompanied by a change in the colour according to the additive in the basic glass mix. Selective absorption of energy in the near infra-red region can reduce solar heat without greatly affecting the transmission of visible radiation. This is in some cases achieved by including iron (ferrous oxide) in the glass which results in a bluish green colour. The addition of nickel and cobalt oxides and selenium gives the grey and bronze heat-absorbing glasses.

There are three principal types of manufacturing processes for flat glass: the *float* process **7** manufacturing the better quality glass with inherently parallel surfaces free from distortion; the *drawn* sheet process **8** now being largely replaced world-wide by the float process; the *rolled* process **9** for patterned and wired glass.

The three processes use basically similar melting furnaces, called tanks, which are designed for large-scale production.

Considerable production time is lost in an initial change from clear glass to absorbing glass production, perhaps two or three days, and the removal of all traces of contaminating mix at the end of the run can take up to three weeks.

There is thus the incentive to turn to other solutions to achieve modifications in the visual and thermal performance of the basic glass product. Since better heat rejection is obtained by improved reflection the principal effort is toward efficient reflecting coatings.

There are three principal types of coated glass achieving reduced transmission characteristics: metallic coatings; optically thin coatings; organic coatings.

Some are durable and used on monolithic glass while others are less so and are protected either by laminating a second piece of glass to the metallic coated glass or by having the coated surface on the inside of a hermetically sealed double glazing unit.

Of these glass processes, double glazing units are principally directed towards thermal and acoustic insulation; lamination is directed towards safety and security against impact, as is toughened glass.

Part I Window Design

1 Window glass design guide

Function 1: provision of view out

1.01 In many cases the prime role of a window is to maintain contact with the world by providing an adequate view. There is wide acceptance of the principle that a view is a desirable feature of the built environment but it is difficult to support the belief with experimental evidence.

Consider importance of view out

1.02 This consideration seems to evoke conflicting responses. Manning[1] found that nearly 90 per cent of respondents to a questionnaire in an office considered that it was important to be able to see out.

1.03 Hardy[2] reports on desirability of a view that less than 13 per cent of his sample of office workers chose 'it does not matter', twice as many chose 'highly desirable'.

1.04 In a recent study based on an acceptance of the window as an essentially desirable feature of buildings, Ludlow[3] reaches a separate conclusion that a view of any quality is better than no view at all.

1.05 Conversely, the small importance of view in relation to other parameters was shown by Markus,[4] in a survey that ranked view among the last four of ten items, and by Hardy,[2] whose study showed that a good view out was judged among the least important of 12 items by 73 per cent of his sample.

Consider window design for view out

1.06 Accepting that a view is to be provided, the primary window design factors will be the geometric considerations of size, shape and position, **1**. The properties of the glazing material are of secondary importance: experience has shown that the visible transmittance of the glass can be reduced to very low levels (less than 0·10) without destroying the view.

1.07 There is little agreement on the necessary shape of the window. Markus[4] makes a case for tall windows that show a vertical slice across the normally horizontal stratification of the outdoor scene, **2**, whereas Keighley[5] shows from model studies that, when the window area is limited, there is a preference for horizontal windows, **3**. Until more research has produced reliable guidance, the details of the window's view and contact function will continue to depend upon the designer's own assessment and preference.

1.08 The quality of the view will be affected by the extent of the sunlight reaching outdoor spaces especially those between buildings. Recommendations on this aspect are discussed in the DOE publication *Sunlight and daylight*.[6]

1

2

3

1 *View out from GLC island block County Hall, London.*
2 *Sixteen per cent of window area shown in* **1** *used vertically.*
3 *Sixteen per cent of window area shown in* **1** *used horizontally.*

Part I Window Design

1 Window glass design guide

Function 2: provision of daylight and sunlight

2.01 It is necessary to distinguish between lighting a task for efficient performance (see next section), and lighting a room for pleasant appearance which is considered here.

Consider daylight amenity

2.02 Many visual tasks are arranged on horizontal surfaces and are best lit by light incident in a generally vertical direction. The appearance of a *space*, however, is more dependent upon the appearance of the vertical surfaces and is therefore influenced mostly by light flowing horizontally. In multi-storey buildings, vertically incident light can, in practice, be supplied only by artificial lighting but horizontal lighting is best supplied by windows, even to considerable distances within deep rooms.

2.03 The combination of vertical artificial light and horizontal daylight is important to the appearance of solid objects within the space. The two can be combined as vector quantities and methods are available for calculating the resultant lighting condition and for assessing its effect on appearance.[7] These rather specialised criteria are fully explained in Chapter 4 and the AJ (11.6.69 p1590 and 9.7.69 p61-64).

1 *Daylight amenity.*

2.04 The strength of the flow of light is denoted by a vector/scalar ratio and Table I tabulates the relation between various values of the ratio and the appearances that they create. There is a general preference for a vector direction between 15° and 45° above the horizontal and the likelihood of achieving this is shown in Table II.

Table I: Vector/scalar criteria for lighting design

Vector/ scalar ratio	Strength of the flow of light	Typical situation	Typical appraisal
3·0	Very strong	Selective spot-lighting. Direct sunlight.	Strong contrasts: detail in shadow is not discernible.
2·5	Strong	Low BZ*, low FFR**, dark floor. Windows on one side, dark surfaces.	Noticably strong directional effect: suitable for display but generally too harsh for human features.
2·0	Moderately strong	Low BZ with medium or light floor. Medium or high BZ. Side windows with light surfaces psali***.	Pleasant appearance of human features for formal or distant communication.
1·5	Moderately weak	Low BZ with medium or light floor. Medium or high BZ. Side windows with light surfaces psali.	Pleasant appearance of human features for informal or close communication.
1·0	Weak	Medium or high BZ with light floor. Side windows in opposite walls.	Soft lighting effect for subdued contrasts.
0·5	Very weak	Luminous ceiling or indirect lighting with light surfaces.	Flat shadow-free lighting: directional effect is not discernible.

* British zonal classification; ** flux fraction ratio; *** permanent supplementary artificial lighting of interiors.

Table II: The percentage of normal working hours (0900-1730) throughout the year for which the vector altitude due to both daylighting and artificial lighting will fall below certain angles*

Sky component on horizontal plane	Average illuminance on horizontal plane due to artificial lighting alone: 200 lux			400 lux			700 lux			1000 lux		
	Maximum vector altitude 45°	60°	75°	45°	60°	75°	45°	60°	75°	45°	60°	75°
1 per cent	73	88	94	46	75	90	5	56	82	0	39	74
2 per cent	87	93	95	73	88	94	53	78	91	30	68	88
3 per cent	91	94	95	82	92	95	70	86	93	55	79	92
4 per cent	93	95	95	87	93	95	76	89	94	66	84	93
5 per cent	94	95	95	90	94	95	81	91	95	73	88	94

* It is assumed that the vector altitude due to daylight alone is not greater than 20° and that the floor is of medium lightness (reflectance 0·2). The prediction of sky components is described in IES Technical Report No 4.[8]

2.05 There are continual changes in the intensity, distribution and colour of daylight and, because all prediction is based on the average state, indoor conditions vary about the design condition. This variability is generally accepted to have a stimulating effect, avoiding the monotony of stable conditions and adding pleasantly acceptable variety.

Consider sunlight amenity

2.06 The appearance of an indoor space will also depend upon the extent to which direct sunlight is admitted. The principal recommendations on sunlighting are given in *British standard code of practice CP3*: 1945[9] which, at the time of writing, is being revised. At present, the aim of the code is to ensure that sufficient sunlight is admitted to dwellings and schools.

2.07 The main recommendation for living rooms is that sunlight should be able to enter for at least one hour each day from February to November. For school classrooms sunlight should be admitted for about two hours each morning throughout the year, see AJ 6.8.69 pp321-325 CI/SfB (E6).

Consider sunlight penetration

2.08 Studies of the penetration of sunlight into buildings are mostly geometric problems that depend upon the relative shapes, sizes and positions of the design point, the room, the window, and the sun.

2.09 The calculations are very suitable for solution by graphic techniques. They rely on knowledge of the altitude and azimuth of the sun at various times of day and at various seasons and, although the apparent movement of the sun in the sky arises from the rotation and orbital progress of the earth, it is convenient and adequate for architectural purposes to consider a stationary earth and a sun that moves across the sky along a path determined by the season of the year and the latitude of the observation point.

2.10 The sun paths can be drawn in a suitable projection and when a sketch of the window, drawn in the same projection, is superimposed, the periods when the sun can be seen from the design point are easily assessed.

2.11 Sets of prepared sun path diagrams are available from several sources: **3** shows the outline of a window on a sun path diagram in the stereographic projection supplied by the Building Research Establishment;[10] **4** shows the same study in the gnomonic (perspective) projection of *Windows and environment*.[11] A variation is shown in **5** in which a gnomonic sun path diagram is used as a horizontal sun dial to help in plotting the visible patches of sunlight. The methods have been fully discussed in AJ 16.10.68 p881-898 (CI/SfB (E6)) and AJ 30.10.68 pp1019-1036 (CI/SfB (E6)).

Consider glass selection for sunlight penetration

2.12 The visual aspects of sunlight are most affected by the size and shape of the window and not much by the properties of the glass.

2.13 The use of low transmission glasses, for example, has very little effect on the sunlit appearance of a room because the patches of sunlight on walls and floor are still evident even if they are a little less bright. Similarly, low transmission glass does not offer a cure for sun glare: if the sun appears within 45° of the direction of view it will cause glare and only the intervention of an opaque shade will give relief.

2 *Sunlight amenity.*

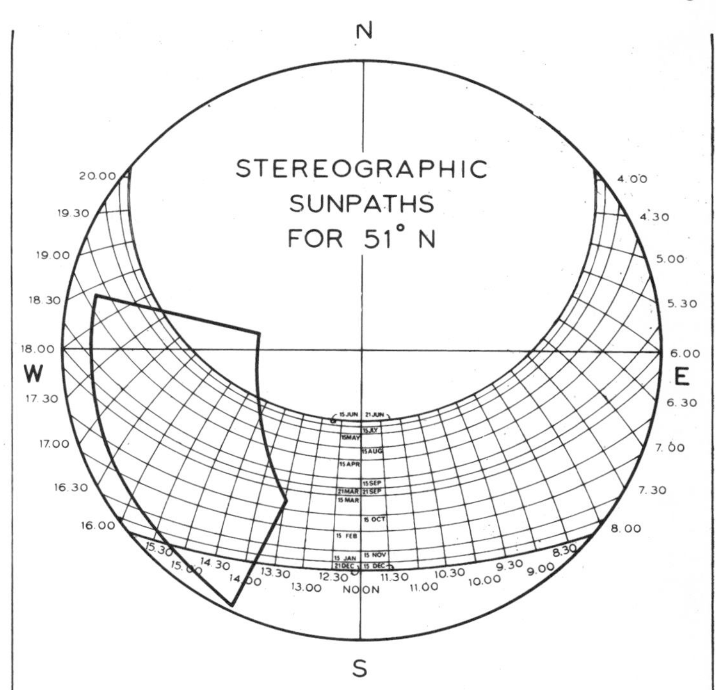

3 *Studying sun penetration using a stereographic projection. Explanation as below, except that window is represented by distorted rectangle resulting from projection onto a hemispherical surface.*

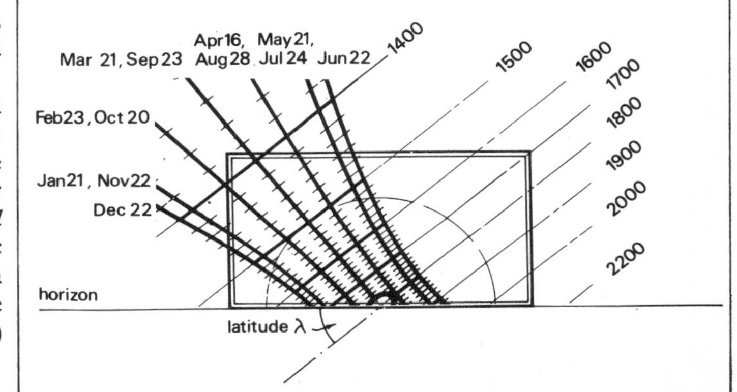

4 *Studying sun penetration using a gnomonic projection. The rectangle is a window seen from a point inside the room: the curves represent the sunpaths on specific dates and the positions occupied by the sun at each time of day are indicated by the intersections with the oblique straight lines.*

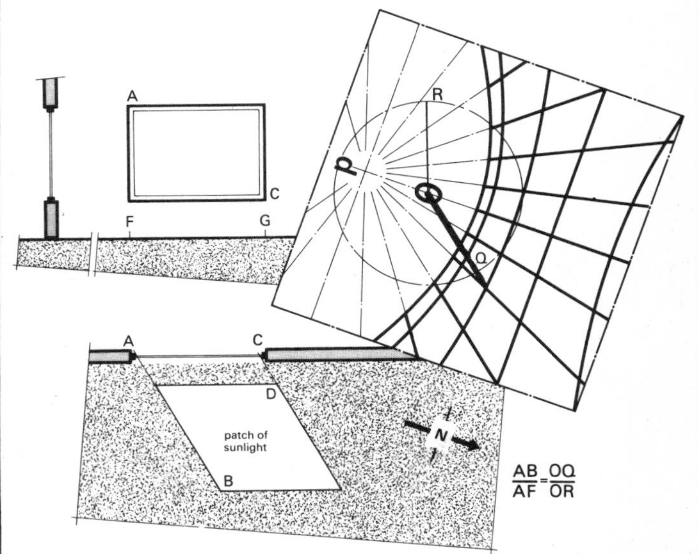

5 *Using a horizontal sun dial to study sun penetration. See AJ* 30.10.68 *pp*1019-1036 *for full explanation.*

Part I Window Design

1 Window glass design guide

Function 3: provision of task illumination

Consider reliance on daylight and integration with artificial illumination

3.01 Daylighting, artificial lighting, and combinations of both can provide the illumination needed for daytime working, the design of each installation being dictated by the tasks to be accommodated.

3.02 Recent lighting design practice has created systems that cover a range from the domestic situation, where daylight is accepted as the daytime illuminant and where intricate tasks that need more light are taken to the window, to the open plan office situation, where daylight is disregarded and high levels of artificial lighting are provided over the whole area of the office.

3.03 The fraction of the total energy input to a building that is used for lighting may, in the domestic situation, be as little as 2 per cent but in modern offices it is commonly more than 50 per cent. It is therefore important to overall fuel economy that daylighting is used to help with task illumination wherever possible.

3.04 Some recent research[12] has shown how, in an office, an artificial lighting installation can be designed to give the right amount of light where it is needed and to make good use of daylight. The energy saved by such a scheme may be 80-85 per cent of that used in a conventional installation. This is discussed in detail in Chapter 6.

Consider daylight prediction

3.05 As in sunlight studies, the important factors in studies of daylighting are those of shape, size and position of window to which is added the distribution of luminance in the sky. These calculations are also very suitable for solution by graphic techniques. Of the many different methods devised, the two that are most used are the sky component protractor method published by the Building Research Establishment,[13] **1**, and the dotted overlay method of *Windows and environment*,[11] **2**.

3.06 It is sometimes necessary to measure daylight factors in existing buildings either as a basis for future alteration or in confirmation of predictions. Examples of the instruments that are used in surveys are the BRS daylight factor meter,[14] **3** which is a direct reading instrument, and the full-field camera with a rapid film processing body that produces a picture of the window on which dotted overlays can be used,[15] **4**.

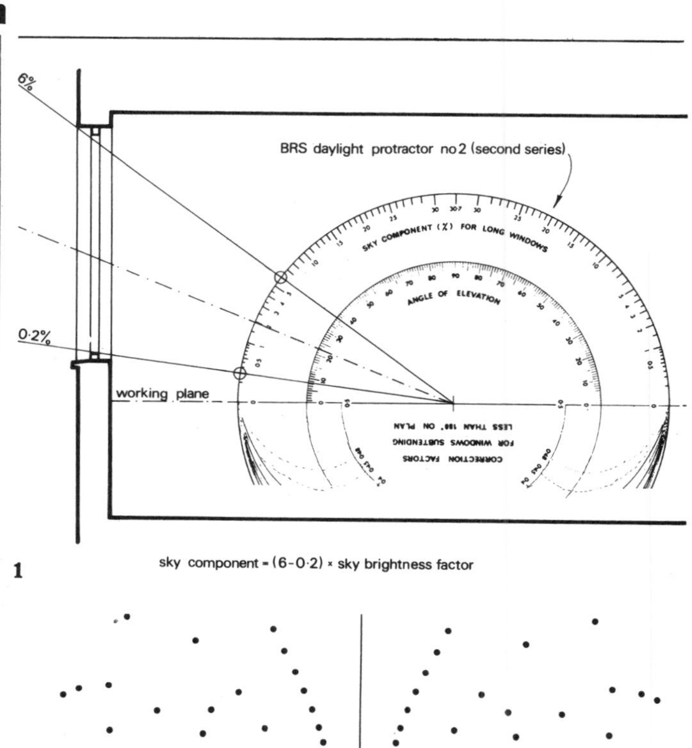

1

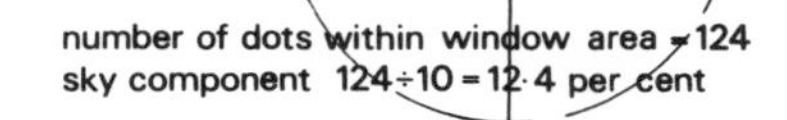

2

3

1 *Estimating sky component by the protractor method.*
2 *Estimating sky component by the dotted overlay method.*
3 *BRS daylight factor meter.*

Consider glass selection for daylight

3.07 The type of glass used in the windows is important to the use of daylighting for task illumination because the daylight factor is directly proportional to the visible transmittance of the glass. This does not mean that the distance that adequate daylight can penetrate into a room is directly proportional to the glass transmittance because the effects of the inverse square law must be taken into account.

3.08 Thus, if a daylight factor of 2 per cent is achieved at a distance of 5 m from a window glazed with clear glass, when the transmittance of the glass is halved, the 2 per cent daylight factor is achieved not at 2·5 m from the window but at about 3·5 m.

Consider daylight recommendations

3.09 Most of the recommendations on levels of daylighting in various situations are based on the provision of adequate light for the accomplishment of the appropriate tasks for a reasonable period (say 85 per cent of the working day).

3.10 The two most important authorities are the code of interior lighting of the Illuminating Engineering Society[16] and the code of practice on daylighting of the British Standards Institution.[17] Some general recommendations are tabulated in Table I. Particular situations may need much more detailed attention than is implied by the generalities in the table.

Table I: Examples of daylight factor recommendations—outlines only

Situation	Recommended daylight factor
church	1 per cent minimum
dwelling—bedroom	0·5 per cent to three-quarter depth
kitchen	2 per cent over half the area
living room	1 per cent to half depth
factory	5 per cent minimum
hospital ward	1 per cent minimum
office	2 per cent minimum
school classroom	2 per cent minimum

4 *Sky component diagram and full-field photograph.*

5 *Task illumination.*

Part I Window Design

1 Window glass design guide

Function 4: to keep out wind and precipitation

Consider factors determining glass strength

Design principles

4.01 Because glass has a considerable variation in strength it has traditionally been designed on a statistical basis.

4.02 Due to the strength depending on a random distribution of flaws, the variation in strength is larger than for most other materials and **1** shows a typical set of test results indicating that for a mean strength of 100 per cent the minimum strength could be as low as 70 per cent whilst the strongest sample of glass could have a strength of 150 per cent of the mean.

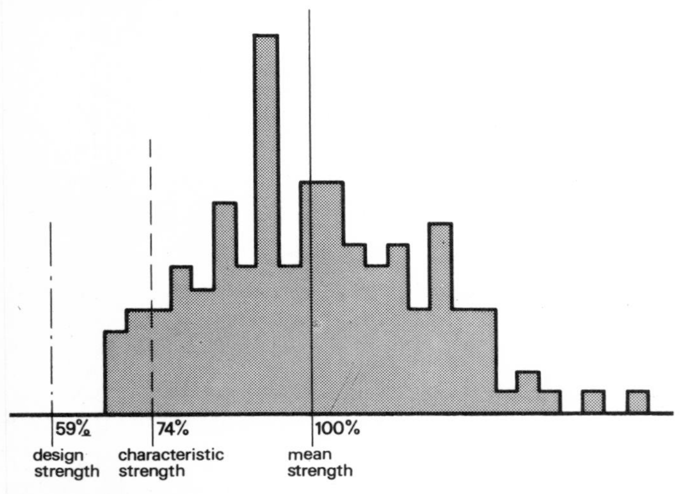

1 *Typical strength distribution for glass. Test results from 145 samples show a wide spread of results around mean value, indicating variance of strength from sample to sample.*

4.03 Because in practice it is often not possible to have a large enough sample for a thorough statistical example, the procedure of statistical tolerance limits has been adopted at the level in which there is 95 per cent confidence, that is, 95 per cent of the samples will have a greater strength.

4.04 The strength value thus obtained is described as the characteristic strength and this is further modified to give a design strength using the principles of limit state design (described for example in BRE publications). The various factors considered and the values given to them are shown in Table I.

Table I Principles of limit state design as applied to glass

The partial safety factor used by Pilkington for glass is 1.25. *It is derived from the multiplication of the following factors:*

Partial safety factor for material strength	1·00
Partial safety factor for constructional errors	1·05
Partial safety factor for exposure effects	1·00
Partial safety factor for brittle materials	1·20

Load duration

4.05 Glass does not exhibit a fatigue process by which it becomes weaker when it is subjected to cyclic loading, but it does demonstrate static fatigue. Static fatigue in glass is the name used to describe the phenomenon by which glass is weaker when subjected to a long duration load than to a short duration load. In practical terms this means that glass can be designed to a higher stress for wind gust loads (usually on a 3 second basis) than for snow loading (considered as being experienced for up to two months per year in the UK).

4.06 The duration of load must be taken into account when glass is being designed and the equations derived give a sufficiently accurate estimate of what occurs, Table II.

Table II Relative breakage stresses for various load durations

Load duration	Example	Relative breakage stress
3 seconds	Wind loading	1·00
1 minute	Some wind loading codes	0·75
Sustained	Snow loading	0·38

Table II shows relative breakage stresses for various load durations; and the use of this table lies in the fact that if, for example, a rooflight was being designed (which may carry a sustained snowload for months), the correction factor of 0·38 shown in the table must be applied to the glass thickness found from charts 2 to 5.

Consider interaction of loadings and glass strength

4.07 The window has to withstand various types of loading:

- Wind loading on the glass (the actual support system is very important eg four edge support, two edge support, flexible supports).
- Other climatic loadings on the glass: 1 snow loading; 2 thermal stressing due to radiation.
- Climatic loadings on the glazing: 1 solar radiation in general; 2 freezing; 3 rain; 4 durability.

4.08 The location and function of the glass may introduce strength limitations which are themselves additional limitations due to:

- Manufacturing processes, ie size limitations on a particular method of glass manufacture.
- Processing limitations, ie the maximum sizes for toughened and laminated glasses or the limited range of glasses which can be made into double glazing.
- Availability limitations, ie tinted glass in the main is only available in 4, 6, 10 and 12 mm thickness.
- Handling limitation: a practical set of limitations on what glass can be used and in what forms.

4.09 The structural limitations due to location and function of glass depend on whether there is:

- A fire requirement (wired glass).
- A safety requirement (often thick and toughened glasses).
- A security requirement (usually laminated glass).
- A requirement for privacy (patterned glasses are weak).

Consider detailed structural design

Design for wind loading

4.10 In order to design glass against wind loading the recommendations of the manufacturers or those given in BS CP 152 'Glazing and fixing of glass for buildings' are used. Reproduced **2**, is one of the four graphs used in the Pilkington catalogue, namely that for transparent glass, ie float, or sheet glass, clear or body coloured when four edge supported. This graph is based on the information shown in Table III and plots the maximum allowable glass area against the design wind loading on a 3 second gust basis, as taken for instance from CP3: Chapter V: Part 2. Lines are shown for the various nominal glass thicknesses available and these are based on the minimum specified thicknesses in BS 952. This ensures that the wind loading recommendations are based on the weakest glass that can be produced and still satisfy BS 952 and reflects the principle used throughout the design of glass against structural loads, of using the 'worst case' in the design.

4.11 Also shown on **2** is a shaded area to take into account the effects on strength of rectangular panes of the same areas as the square pane which is shown by the thick line for each glass thickness. This shaded area is based on the approximate effects of length/width ratio. In special cases detailed calculations can be carried out using the precise theory, but in general the shaded area gives a satisfactory interpretation of what actually happens. The upper limit of the shaded area represents the additional strength of a pane whose shape is such that the longer side is three times the shorter side, ie length/width ratio 3:1. By linear interpolation the strength of a pane of aspect ratio nearer to square (ie 1:1) can be determined. For example, panes of aspect ratio 2:1 are represented by the centres of the shaded areas.

Table III Design stresses used by Pilkington for 3 second load duration on normal windows

Type of glazing	Design stress MN/m²
Transparent glass, 3-6 mm thick	41·0
Transparent glass, 10-25 mm thick	28·0
Double glazing	As for transparent glass
Laminated glass	As for transparent glass
Patterned glass	27·3
Wired glass	20·9
Rough cast glass	25·0

4.12 For glass subjected to wind loading and supported on two opposite edges only or on three edges, similar graphs can be produced using the general engineering formulae for such conditions. In both these cases details of the actual support system are important in order to determine whether the face strength of the glass should be used or its edge strength. The edge strength will vary with the particular process, even the cutting method used on the glass, and the manufacturer should be consulted for strength recommendations.

4.13 If in addition the area of each glass pane is known it is possible from **2** to calculate what glass thickness is required for the length/width ratio. In practice, rather than trying to use the graph in two directions at the same time, it is often easier to use the graph to determine what wind loadings, pressure or suction, glass plates of particular areas, aspect ratio and thickness will stand and from these determine what glass thickness is needed to meet the particular wind loading.

4.14 Similar graphs are shown for:

- Rough cast glass, patterned glasses and wired glass (both polished wired and rough cast wired glass), **3**.
- Pilkington double glazing units in which both glasses are the same strength and thickness, **4**.
- Laminated glasses with a 0·38 mm thick interlayer of polyvinyl butyral, **5**.

For other glass types, the manufacturer should be consulted when the wind loading strength is required.

Method

4.15 The following method can be used for calculating required glass thickness in order to withstand wind pressure, where sheets of glass are large. It makes use of CP3: Chapter V: Part 2: 1972 *Wind loads,* a standard reference which ought to be on architects' shelves and the relevant parts of which are too long to reproduce here.

- Determine *basic wind speed* in metres/second for the particular location of the building under consideration (from section 5.2 of the above Code of Practice).
- Deduce the *design wind speed* from the above, by applying to the basic wind speed three adjustment factors given in section 5.3 of the CP—S_1 which takes account of the effect of hills and valleys in the vicinity; S_2 which takes account of ground roughness, building size and height above ground; and S_3 which takes account of the number of years over which there will be exposure to the wind.
- Convert the design wind speed into a *dynamic pressure* in N/m² by using Table IV in the above CP.
- By locating the latter value on the horizontal axis of the appropriate chart **2** to **5** on page 8, it is now possible to read off the required glass thickness for any proposed glass area. If, for instance the design wind speed was 59 m/s, table IV of the CP will indicate that the dynamic pressure is 1470 N/m²; and chart **2** in turn will indicate (as shown by the broken line) that glass thickness for 1·4 m² pane area ought to be 3·7 mm in the case of clear glass. In fact, the next higher commercially available thickness will be used. For an explanation of which diagonal to use on **2** (the solid line, the shaded band, or the white band) see paragraph **4.11**.

Design for snow loading and self-weight of glass

4.16 Whenever glass can be used on the outside of a building other than for vertical glazing it will be necessary to consider the combined effects of snow loading and the self-weight of the glass as these can provide a further design loading condition.

4.17 Glass demonstrates static fatigue and in practical terms this means that glass is weaker (ie it should be designed to a lower stress) when subjected to snow and self-load.

Snow loading. This is specified in CP3: Chapter V: Part 1 according to the angle of the glass measured from the horizontal and can be translated to a sustained loading normal to the surface of the glass as shown in table IV. This table is based strictly on CP3: Chapter V: Part 1 which describes the snow loading in terms of the loading 'on plan' which results in a cosine squared term in the conversion to the loading normal to the glass.

4.18 If such sustained loads are to be considered in conjunction with wind loadings (which are on a 3 second basis) then it is necessary to multiply the snow loadings by a factor so that their effect on the strength of a glass can be considered in terms of wind loading.

Table IV Snow loading

Design snow loads perpendicular to the glazing

Angle of glazing	*Load N/m²*
0	750
5	744
10	727
15	700
20	662
25	616
30	562
35	447
40	342
45	250
50	172
55	110
60	62
65	30
70	10
75	0

To accord with CP3: Chapter V: Part 1, the design snow load perpendicular to the glazing should be taken as 750 $\cos^2 A$ up to 30° roof slope and as 750 $\left(1-\frac{(A-30)}{45}\right)\cos^2 A$ from 30° up to 75°. At 75° by the second expression, the snow load is zero and it remains zero up to 90°.

CLEAR GLASS (except Laminated)
(3 second mean wind loadings)

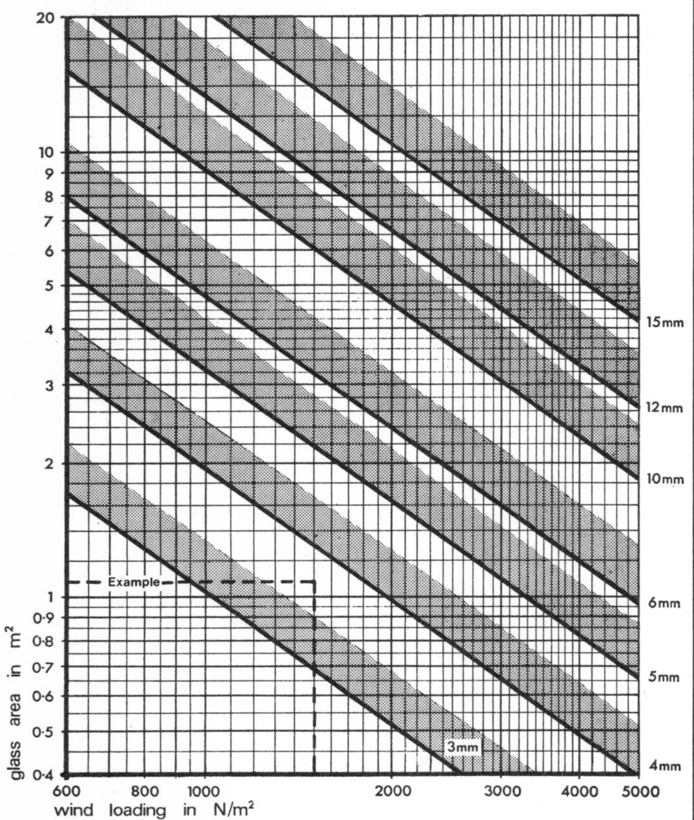

2 *If wind loading is known (see paragraph* **4.15** *for calculation method), suitable glass thickness can be found for required pane area. Eg a loading of 1500 N/m² if applied to a pane of 1·4 m², will require a glass thickness of at least 3·7 mm.*

WIRED, ROUGH CAST AND PATTERNED GLASS
(3 second mean wind loadings)

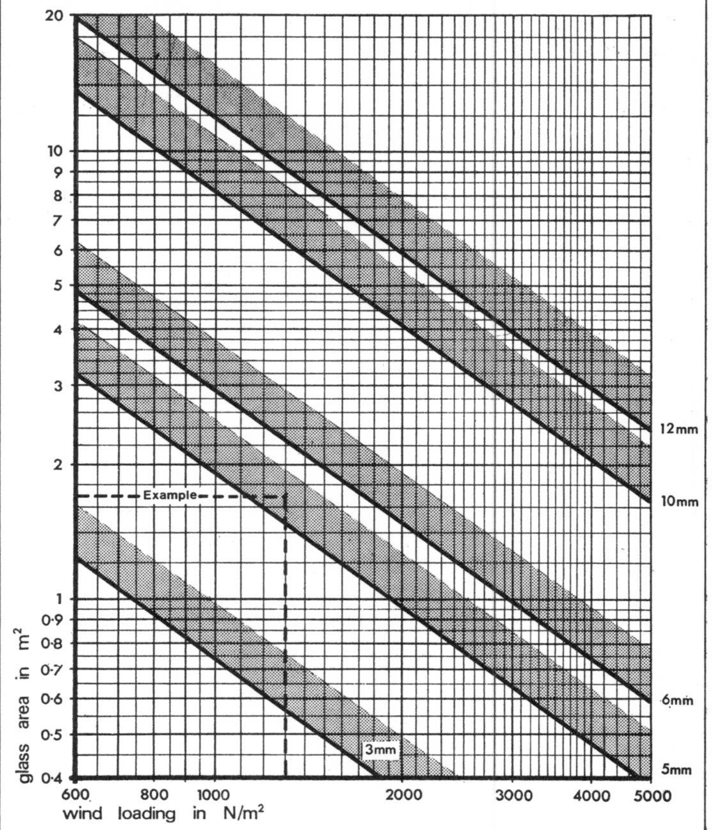

3 *Similar chart to* **2,** *for wired, rough cast and patterned glass.*

DOUBLE GLAZING (CLEAR)
(3 second mean wind loadings)

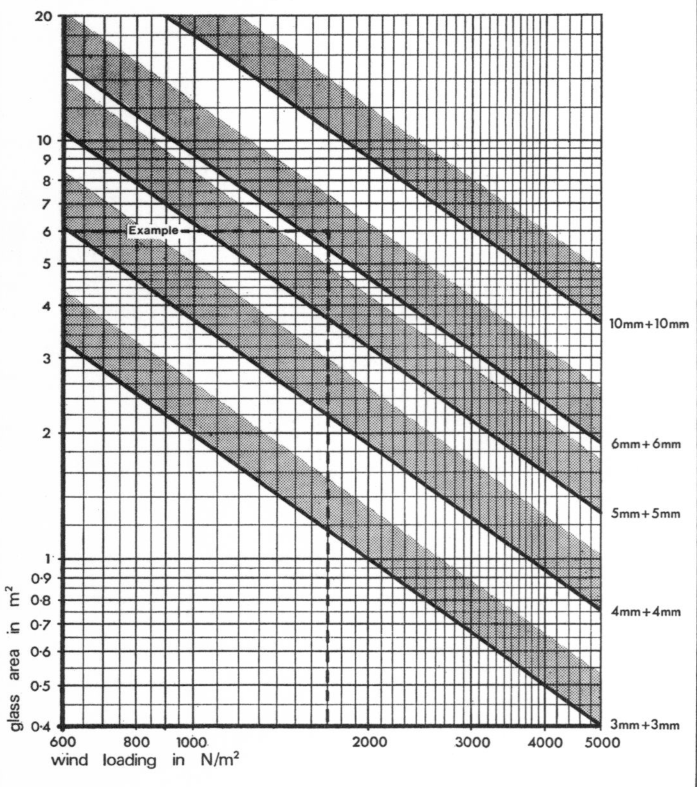

4 *Similar chart to* **2,** *for clear double-glazing.*

LAMINATED GLASS
(3 second mean wind loadings)

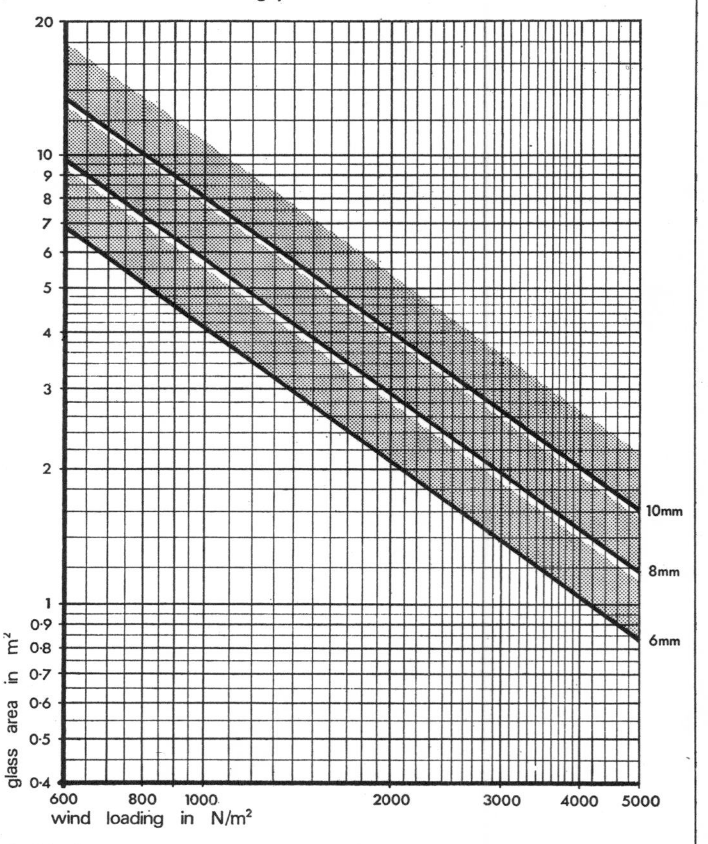

5 *Similar chart to* **2,** *for laminated glass.*

4.19 In practice the combination of stresses is calculated using the following equations:

$P_{press} = + p_{wind} + 2{\cdot}6\,(p_{snow} + p_{glass\ weight})$

$P_{suc} = - p_{wind} - p_{glass\ weight}$

for wind pressure and suction respectively. The factor of 2·6 converts long sustained loads to a wind loading basis and corresponds to a snow loading existing for 56 days. In the second equation, because the glass weight counters the wind suction only when the suction occurs, the correction factor is not applied to the glass weight.

Consider rain exclusion

Design for glazing seals

4.20 Standard structural construction is designed for rain falling at 45° to the vertical. In high winds, however, rain can strike the face of a high building over an arc as wide as 180°, ie it may strike downwards, horizontally or upwards. In Britain the water run off could collect about 14 litres/m run/floor/minute during a heavy storm, **6**.

6 *Rain penetration.*

4.21 It is clear that the majority of windows are glazed by relatively traditional methods under site conditions which can sometimes leave much to be desired.

4.22 An intermediate group of sealants includes butyls, acrylics and polythene, chloroprene and vinyl formulations. Generally speaking they have better weathering and adhesion characteristics but still cannot absorb much movement. The relative cost is perhaps double that of the low performance materials.

4.23 Finally, the relatively high cost, high performance, glazing seals which include the polysulphides, silicones, polyurethanes and solvent-release type acrylics all perform very well with a much greater degree of movement in the joint. They also resist exposure to the weather at least twice as well as the sealants in the intermediate group but considerably more care must be taken in handling and applying and their cost ranges from three to six times that of the materials in the intermediate group. Even so the cost of these high performance sealants is infinitesimal compared with the total cost of a building and there is surely no better place to make a good investment than in good glazing seals, **7**.

4.24 The comparatively new technique of open drained glazing joints lends itself particularly to dry gasket glazing but once again, proper design based on adequate practical experience is necessary to avoid the pitfalls in this matter. All too often

7 *A selection of glazing materials.*

drainage holes or slots of inadequate size lead to eventual blocking with consequent leakage into the building or attack on the edges of laminated glasses or sealed double glazing units by the entrapped contaminated water, **8**.

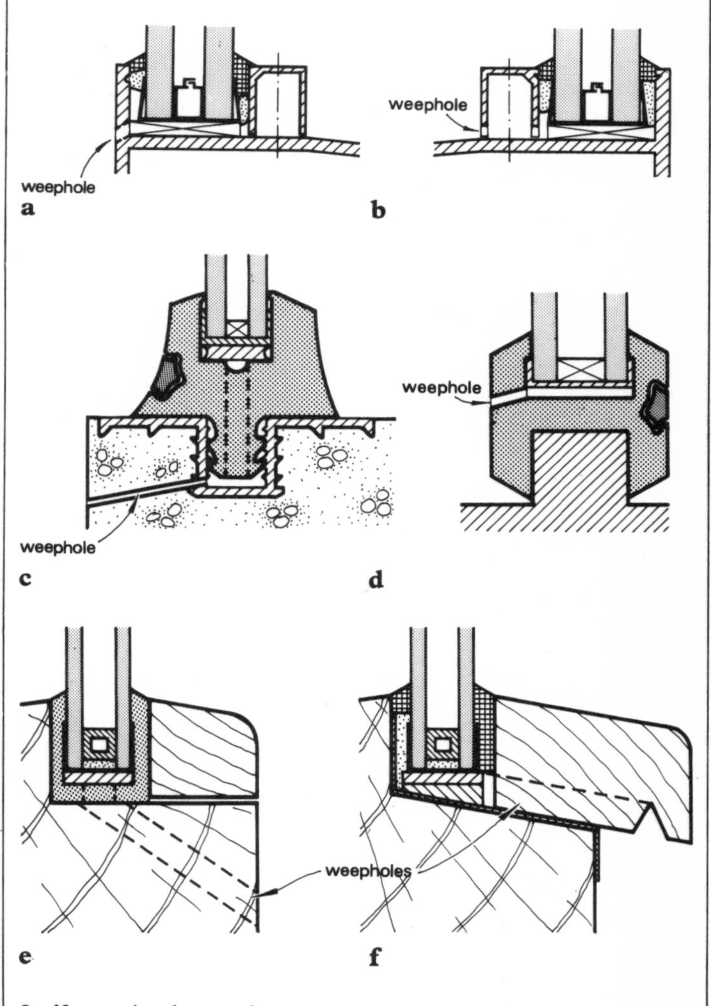

8 *Alternative forms of drained glazing: metal frame* **a, b;** *structural gasket* **c, d;** *timber frames: glazing with gasket* **e;** *mastics* **f.**

Part I Window Design

1 Window glass design guide

Function 5: as a light filter

Consider the effects of low light transmittance glasses
5.01 Some of the results of the light-filtering functions of windows have been discussed in sections two and three. Qualitatively, the effects of low transmittance of the glass are most significant to *task* lighting and it also reduces the amount of daylight available as a horizontal component in *lighting quality*. Quantitatively, low transmittance reduces the *glare* from a bright outdoor scene. Otherwise, the effects are small.

Consider the effects of coloured glasses
5.02 The colour effects of tinted glasses are less than might be expected. Because of the efficiency of adaptation of the eye it is surprisingly difficult to identify the hue or amount of tint when there is no contrasting view (either within the space or on point of entry) through clear glazing to provide a reference.*
5.03 Some care in use of tinted glass is required for situations where intricate colour matching is necessary. But there is evidence[18] that colour discrimination is less affected by tinted glasses than by colour of artificial light sources.
5.04 A study[19] of the use of tinted glass in the design of offices has shown that where reduced transmittance glasses are used, there is a preference for the *bronze* types for their appreciably warm tint. Other tints may be used satisfactorily if the colour saturation is low. A warm-tinted glass is best used in conjunction with an interior colour scheme that is either intermediate or warm in tone (ie not cool) and with 'warm white' fluorescent lighting (or similar source with a correlated colour temperature of less than 4000 K). A grey or neutral tinted glass is more acceptable in conjunction with a cool interior colour scheme and electric lighting of relatively high colour temperature.

* See AJ 4.6.69 pp1527-1544 for general background reading.

1 *Anti-sun bronzefloat glass used in College of Domestic Science Building, Glasgow by Building Design Partnership, architects.*
2 *Bronze toughened glass used in offices for Willis, Faber & Dumas Ltd in Ipswich by Foster Associates, architects.*
3 *Bronze solarshield glass used in Herringthorpe leisure centre, Rotherham, Yorkshire.*
4 *View out from the Polar building, Tokyo by Nikkon Sekkei.*

1

2

3

4

5

Consider the internal effects of reflective glasses

5.05 Some solar control glasses and window treatments work by increasing the surface reflectance and the resultant visual effects must be taken into account.

5.06 The subjective effects of reflections in the glass for the *outward* viewing situation formed part of the Newcastle study* which concluded that the more often reflections occur the more likely they are to be considered pleasant; but that although the reflections themselves attract little adverse comment, they do tend to interfere slightly with the enjoyment of the view.

5.07 The occurrence of reflections depends upon brightness of reflected view in relation to that of transmitted view.

5.08 There is no such thing as a one-way-vision glass, only a one-way-vision situation: interchanging the brightnesses of the scenes on the two sides of the glass always reverses the direction of one-way-vision. Similarly, because clear glass has a reflectance of 0·08, reflections can occur in clear glass but only when the difference between the brightnesses of the two scenes is greater than for reflective glass.

5.09 In physical terms the difference between clear and reflective glass therefore is one of degree only. But there is a practical difference in that, unlike clear glass, reflective glass tends to be used without curtains that would be closed at dusk and mask the reflections. The reflections may give a sense of privacy at night by masking all view of the outdoor scene; but this privacy would be illusory because the inward view is not impeded when the interior is lighted and the outdoor scene is dark.

Consider the external effects of reflective glasses

5.10 The outdoor effects of reflective glasses, apart from the mirror-like appearance of the building, are again only different from those of clear glass in degree. Any effects such as glare (due to the reflection of the sun or the headlights of a car) pose geometric problems.

5.11 The solutions lie in changing the direction of the reflected light or interposing an opaque barrier. This is because, in reflective glasses, the reflectance is increased by a factor of no more than 6 and (just as direct sun glare through a window cannot be cured by low glass transmittance unless the transmittance is very much less than 0·05), a change from reflective glass to clear glass would not reduce the brightness of a reflection by enough to cure glare.

5.12 There may, occasionally, be significant problems in particular cases where the solar energy reflected from a reflective facade provides an unacceptable cooling load on a facade.

* AJ 8.5.74 pages 1035 to 1040.

6

7

8

9

5 *External view by day of Prospect House, Princeton, New Jersey by Warren Platner, architect.*
6 *External view by night of the same building.*
7, 8 *External views of reflective glass used in Norgas House, Killingworth by Ryder & Yates, architects.*
9 *Internal view of reflective glass in the same building.*

1 Window glass design guide

Function 6: as a sound filter

6.01 With increasing road traffic densities and the advent of noisy aircraft and construction machinery there are many places where outdoor noise levels are annoyingly high. Such situations impose an extra restraint on window design. Because glass is usually used in such thin layers, single glazed windows rarely carry out their lighting and view functions without being a weakness in the building's defence against noise.

6.02 The first principle that applies in improving sound insulation is to block all gaps and cracks—they transmit much more sound than their fractional area would suggest. So, good sound insulation in very noisy environments requires sealed facades, but that precludes natural ventilation and demands artificial means of removing the heat generated within buildings and of avoiding overheating from solar heat gain.

6.03 For any material the sound insulation of a panel can be improved by increasing the thickness. Thus, the effect of doubling the thickness of glass is to increase the average sound insulation by a little over 3 decibels which is probably a worthwhile improvement with the thinner glasses but soon ceases to be economical. It is better to use the extra glass as a second leaf in double glazing.

6.04 As the air space of double glazing is made wider the sound insulation is increased, although the magnitude of the increase depends upon the frequency of the sound and the properties of the air space. Thus, some improvement in sound insulation can be achieved by lining the perimeter of the air space with acoustically absorbent material.

6.05 The highest values of sound insulation can be obtained in a construction comprising double windows in separate frames set in separate surrounds: this form is usually restricted to situations where the noise problem is particularly severe or to other special applications where the expense is justified.

6.06 Sealed double-glazing units have the disadvantage that the air space cannot be made very wide and the panes are mechanically coupled at the edges. A narrow air space double-glazing unit which would be thermally satisfactory would not be recommended for low frequency noise reduction but the converse is not true—any double glazed window designed to have good sound insulation will also have good thermal insulation.

6.07 Sound insulation values are very dependent upon the frequency of the sound. The curves in **1** are representative and show the general increase with increasing frequency as well as some of the dips that can be caused by resonances within the glass or within the air space.

6.08 The sound insulations of a selection of single and double glazings are listed in Tables I and II. The average values quoted in the tables are for the range of frequencies from 100 to

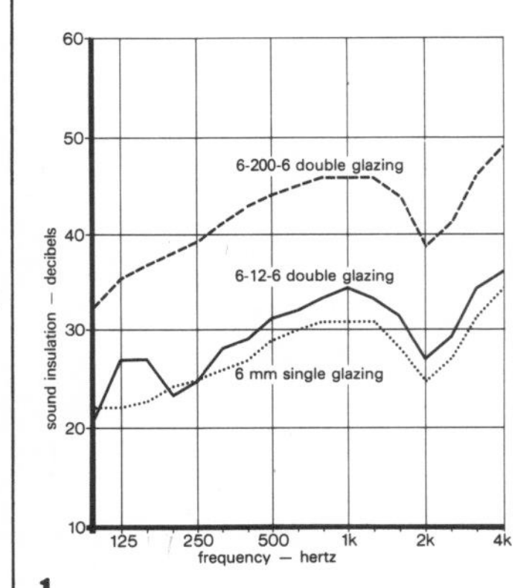

1 *The effect of width of air space on sound insulation. The three numbers refer to thicknesses of glass sheets and air space between.*

Table I Sound insulation of single glazing (dB)

Centre frequency of third-octave band* (Hz)	Glass thickness (mm) 4	6	8	10	12
100	20	22	23	24	25
125	20	22	24	25	26
160	21	23	25	26	27
200	22	24	26	27	28
250	23	25	27	28	29
315	24	26	28	29	31
400	25	27	29	30	32
500	26	29	30	31	33
630	27	30	31	32	33
800	28	31	32	32	32
1000	29	31	32	31	29
1250	30	31	29	27	26
1600	31	28	26	27	30
2000	29	25	28	31	34
2500	26	27	32	35	37
3150	23	31	35	37	40
4000	28	34	38	40	42
Mean (100-3150 Hz)	25	27	29	30	31

Table II Sound insulation of double glazing (dB)

Centre frequency of third-octave band*	Thickness of glass/air space/glass (mm) 6/12/6	10/12/6	10/100/6	10/200/6	6/200/6	6/200/6
(Hz)	Sound absorbent in reveals No	No	Yes	Yes	Yes	No
100	21	22	26	33	32	32
125	27	28	29	37	35	35
160	27	28	32	39	38	37
200	23	24	34	41	40	38
250	25	25	36	42	42	39
315	28	29	38	44	43	41
400	29	30	40	45	45	43
500	31	31	42	47	46	44
630	32	32	44	48	47	45
800	33	33	45	49	48	46
1000	34	34	47	50	49	46
1250	33	34	47	50	50	46
1600	31	33	48	50	49	44
2000	27	34	50	51	44	39
2500	29	35	52	52	47	41
3150	34	36	54	54	52	46
4000	36	38	56	56	55	49
Mean (100-3150 Hz)	29	31	42	46	44	41

* See paragraph **6·12**

3150 Hz, in line with European practice. In America the value for 4000 Hz is included which, because it is usually the highest, explains why higher averages are quoted in America.

The response of the human ear varies with frequency

6.09 For a given sound pressure level a low frequency sound does not appear to be as loud as a sound of the same sound pressure level at a higher frequency.

6.10 For example a pure tone with a sound pressure level of 85 dB at 100 Hz sounds no louder than 70 dB at 4000 Hz. To make allowance for this non-linear characteristic of the ear, instruments for measuring noise are fitted with 'weighting' networks which selectively attenuate the meter's response.[20] The best known is the 'A' weighting network. Its response is shown in **2**. Recent surveys have shown a good correlation between subjective assessment and sound levels measured in dBA and this has led to the extensive use of dBA.

Glass selection for a noise problem

6.11 To assess a noise problem in more detail than the measurement of an overall noise level allows, use can be made of filters that transmit energy at all frequencies in a given band and attenuate at all other frequencies.

6.12 These band-pass filters can be designed to pass noise in any width of band centred on any frequency but, to facilitate comparison of measurements, standard centre frequencies and band widths have been defined.[21] The centre frequencies of the third-octave bands are those listed in Tables I and II and the standard centre frequencies for the octave bands are 125, 250, 500, 1000, 2000, 4000 Hz.

6.13 Although dBA is now widely accepted as an appropriate measure, in many noise investigations when sound insulation is being studied it is preferable to have the additional information that band analysis can provide. The method is illustrated in **3**.

6.14 At each frequency, the insulation of the proposed barrier is subtracted from the sound level in the outdoor noise spectrum to give the indoor sound level. In this way the indoor noise spectrum can be plotted, compared with a set of Noise Rating curves **4** and assigned a Noise Rating number derived from the lowest curve that is not exceeded by the noise spectrum—NR45 in the example.[22]

6.15 The Noise Rating curves, in conjunction with recommended maximum Noise Rating numbers and corrections for location and character of noise, have been proposed as an international standard. See Table III.

Table III Recommended maximum noise rating numbers (NR)

Situation	NR
Class room, conference room (50 seats)	25
Private office	40
General office	50
Office (with typewriters)	55
Workshop	65
Sleeping room (see corrections below)	25
Living room (see corrections below)	30
Corrections for dwellings*	
(a) Pure tone easily perceptible	− 5
(b) Impulsive and/or intermittent	− 5
(c) Noise only during working hours	+ 5
(d) Noise during 25 per cent	+ 5
6 do	+ 10
1·5 do	+ 15
0·5 do	+ 20
0·1 do	+ 25
0·02 do	+ 30
(e) Very quiet suburban	− 5
Suburban	0
Residential urban	+ 5
Urban near some industry	+ 10
Area of heavy industry	+ 15

* In the situations described, reduce or increase the choice of NR as indicated.

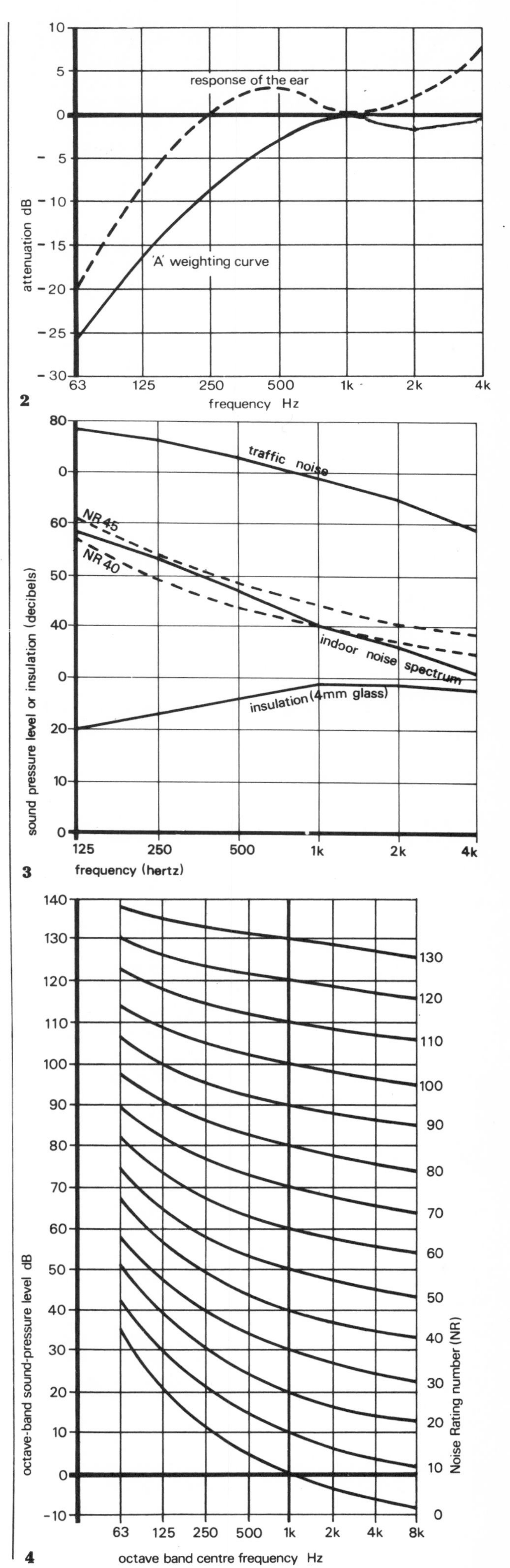

2 *The 'A' weighting curve applied to sound pressure measurements corresponds well with the response of the ear.*

3 *Estimating the indoor noise rating. For typical traffic noise curves see AJ 29.1.69 pp335-341.*

4 *Noise rating curves and the NR numbers applied to them.*

6.16 The foregoing comment that sound levels measured in dBA show good correlation with subjective assessments is especially true for road traffic noise. The *London noise survey*[23] is still the source of reliable information about urban noise levels. They are related to locality, Table IV, because the recording units were left unattended and there were no observations of traffic quantity, composition or speed. Table V gives acceptable indoor noise levels in dBA[24]. If these are subtracted from the values in Table IV the differences lead to the required average insulation values for windows given in table VI. General recommendations on the type of window to suit various locations are given in Table VII.

6.17 Recommendations like those in Table VII are useful for general guidance and initial decision-making but they rely on simplifications and take no account of such influences as frequency effects, flanking transmissions and the modifications caused by reverberation within the room. The final selection of window type is best made on the basis of a detailed analysis of the window in relation to the noise spectrum, **3**.

Consider the legal aspects of sound insulation

6.18 Until the Control of Pollution Act can be implemented and apart from the general law on nuisance, the law relating to the sound insulation properties of windows is only concerned with aircraft noise and road traffic noise and their effects on homes. In both cases the spur to improving sound insulation is in the form of a grant to cover the cost of improvement (see AJ 9.7.75 p63).

6.19 Against aircraft noise the grant scheme is administered by the airport authority at certain airfields, notably Heathrow, Gatwick, Luton and Manchester. Highway authorities must offer the cost of insulating homes against increased traffic noise that arises from the building of new roads or alteration to old ones.

Aural assessments of design solutions

6.20 One of the difficulties of the design of windows for sound insulation is that there is normally very little opportunity to make subjective assessments before the building is complete. An instrument that can help with the problem, the Pilkington Spectrum Shaper, contains sixteen third-octave by-pass filters with centre frequencies ranging from 100-3150 Hz as listed in Tables I and II.

6.21 The attenuation of each of these filters can be changed in step of 2 decibels up to 70 decibels so that the sound insulation characteristics of any practical window can be matched. The spectrum shaper is then used to modify a tape-recording of the incident noise in precisely the same way as the proposed window would modify the real noise. This allows the observer to make his own aural assessment of the effect.

Table IV Noise levels exceeded for 10 per cent of the time at various locations

Group	Location	10 per cent Noise levels—dBA 0800-1800 hours	0100-0600 hours
A	Arterial roads with many heavy vehicles and buses (kerbside)	80	68
B	(i) Major roads with heavy traffic and buses (ii) Side roads within 15-20 m of A or B (i) groups roads	75	61
C	(i) Main residential roads (ii) Side roads within 20-50 m of heavy traffic routes (iii) Courtyards of blocks of flats screened from direct view of heavy traffic	70	54
D	Residential roads with local traffic only	65	52
E	(i) Minor roads (ii) Gardens of houses with traffic routes more than 100 m distant	60	48
F	Parks, courtyards, gardens in residential areas well away from traffic routes	55	46
G	Places of few local noises and only very distant traffic noise	50	43

Table V Acceptable noise levels for various situations

Situation	dBA
Large rooms for speech such as lecture theatres, large conference rooms or council chambers	30
Bedroom in urban area	35
Living room in country area	40
in suburban area	45
in busy urban area	50
School classroom	45
Executive office	45-50
General office	55-60

Table VI The required sound insulation of windows for various reduction of road traffic noise

Difference between outdoor and indoor levels—dBA	Window insulation (100-3150 Hz)—dB*
20-25	20-25
28	30
30	35
35	40
40	45

* See last row in tables I and II.

Table VII Windows

Situation	Location* A/B	C	D	E	F/G
	Type of window				
Lecture theatre etc		Double	Double	Sealed heavy	Sealed light
Bedroom	Double	Double	Sealed heavy	Sealed light	Openable light
Living	Sealed heavy	Sealed light	Openable light	Openable light	Openable light
Classroom	Double	Sealed light	Openable light	Openable light	Open
Executive office	Sealed heavy	Sealed light	Openable light	Openable light	Open
General office	Sealed light	Openable light	Open	Open	Open

* See table IV.

Part I Window Design

1 Window glass design guide

Function 7: as a heat filter

Heat gain

Consider the incident solar heat on the glass and get an idea of the solar energy that is striking the glass.

7.01 As for sunlight penetration, the admission of all solar radiation (heat as well as light) is a function of the relative sizes, shapes and positions of the building, the window and the sun; of the properties of the glass; and of the radiation intensity.

7.02 Graphic methods are useful in understanding and manipulating the solar energy on the façade. Most methods used are a variant of the Waldram diagram,[25] **1**, the Building Research Establishment overlays, **2**, and the *Windows and environment* overlays.

Consider the transmission of solar heat through the glass

7.03 When solar radiation is incident upon glass some is reflected back from the surfaces, some is absorbed in the glass and the remainder is transmitted through, the exact proportions depending on the properties of the chosen glass. An example, for a heat-absorbing glass is shown in **3**.

7.04 The energy that is absorbed (49 per cent in the example) serves to raise the temperature of the glass until thermal equilibrium is reached. After this point the energy is not retained in the glass but is dissipated to the air on the two sides in proportions that, for a given glazing system, are mostly dependent upon the exposure, ie upon the outdoor wind speed.

7.05 Three degrees of exposure, 'sheltered', 'normal' and 'severe' are defined by the Institution of Heating and Ventilating Engineers.[26] It is usual to compare structures under 'normal' exposure conditions, ie for a wind speed of 2 m/s. For the glass in **3**, the partition factor (the fraction of the absorbed energy that is admitted) for normal exposure is 0·31 so the admitted part of the absorbed energy (0·31 × 49) is 15 per cent and this must be added to the directly transmitted 45 per cent to give the instantaneous total transmittance, 0·60.

7.06 The two modes of heat admission are considered separately because they differ in their interactions with the environment. The heat from the absorbed energy is admitted by its heating effect on the indoor air which is immediately apparent and applies an immediate load on any air conditioning plant.

7.07 The directly transmitted radiation, however, must be absorbed by some indoor surface thus raising the surface temperature before heating the air. There is therefore a time lag before a load is imposed on air conditioning plant and the eventual load is reduced by a storage factor that depends upon the thermal capacity of the structure. A selection of storage factors and time lags is tabulated in table I.

Glass selection and solar heat gain

7.08 There is now a wide variety of glasses available with

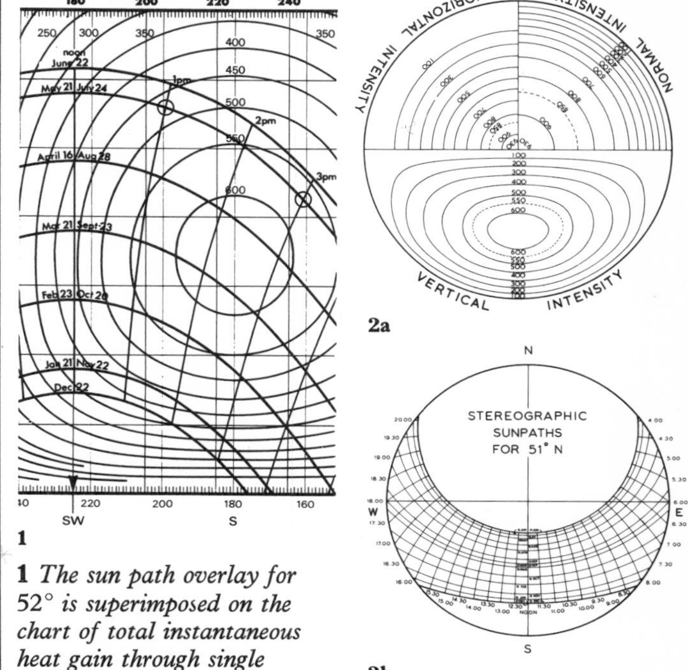

1

1 *The sun path overlay for 52° is superimposed on the chart of total instantaneous heat gain through single glazing. The window faces SW. At 1 pm 24 July the heat gain is 473 W/m². The peak gain is seen to occur at 3 pm—570 W/m².*

2a

2b

2a *The overlay, showing direct solar radiation in metric units.* **2b** *The BRS stereographic projection.*

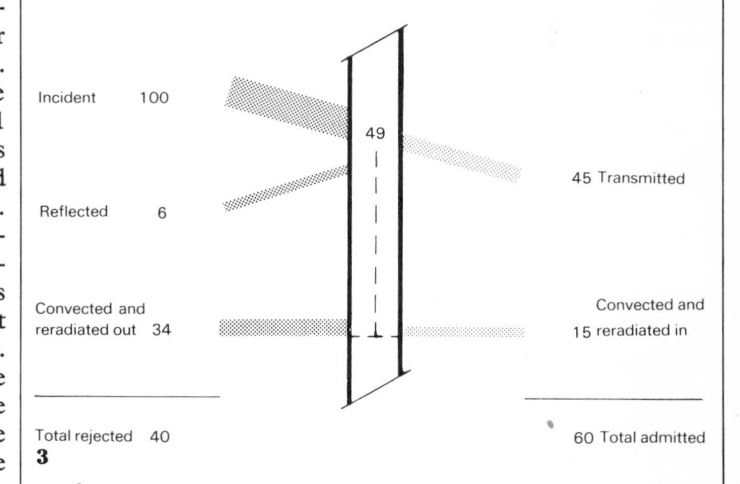

3

3 *The distribution of solar radiation by a heat-absorbing glass.*

properties in the ranges tabulated in table II. These ranges are not independent because, for example, the sum of reflectance, absorptance and transmittance must always be unity.

7.09 The shading coefficient, in which the total instantaneous transmittance of the glass is expressed as a fraction of 0·87 (the transmittance of a notional clear glass between 3 and 4 mm thick), gives a convenient method of comparing the performances of different glasses. It takes no account of the effects of thermal storage in the building but, in general, the lower the shading coefficient the more heat is rejected by the window.

7.10 Increased reflection and increased absorption are both used by solar control glasses as techniques for controlling solar heat gain but one mode is generally predominant and the glass is classified as 'reflective' or 'absorptive'.

7.11 *Reflective glasses* are made by modifying the surface of the glass during manufacture or by depositing on the glass surface a reflective layer, either metallic or dielectric. Similar effects can be achieved by the later addition of a reflective plastic layer. The rejection of solar radiation by reflection can also be increased if the glass is tilted so as to increase the angle of incidence: the effect on the transmittance of clear glass is shown in table III.

Absorptive glasses contain small amounts of materials that colour the glass so that selected parts of the solar spectrum are absorbed throughout the thickness of the glass.

The resulting rise in temperature of the glass causes the glass to expand and sets up stresses that can occasionally become great enough to fracture the glass. Good design practice and careful adherence to recommended glazing procedures avoid these thermal fractures.

Consider further modification by the alteration of window area or by the use of shading devices

7.12 For those situations where sun glare or the thermal discomfort of direct sunshine can be alleviated only by interposing a shading device the design techniques shown on page 3 (Function 2) can be used. **4** exemplifies the use of the gnomonic projection in designing a shading device to exclude the sun from the design point during certain periods. The adoption of a criterion, such as that sunshine shall not be admitted from March to September, establishes the position of the edge of the shade from which the dimensions of a fixed blind or a canopy can be derived.

7.13 Shading devices affect the solar heat gain through windows. Canopies and other external shades reduce the instantaneous heat gain. The energy that they absorb is almost entirely dissipated to the outdoor air.

7.14 Blinds and louvered shades that form part of the window system lose some of their absorbed energy to the indoor air and the thermal properties of the window as a whole depend upon the type of glass, the type of glazing, the position of the blind and the transmissive and reflective properties of the blind material. Shading coefficients for a selection of window designs are tabulated in Table IV, taken from Chapter 8.[27]

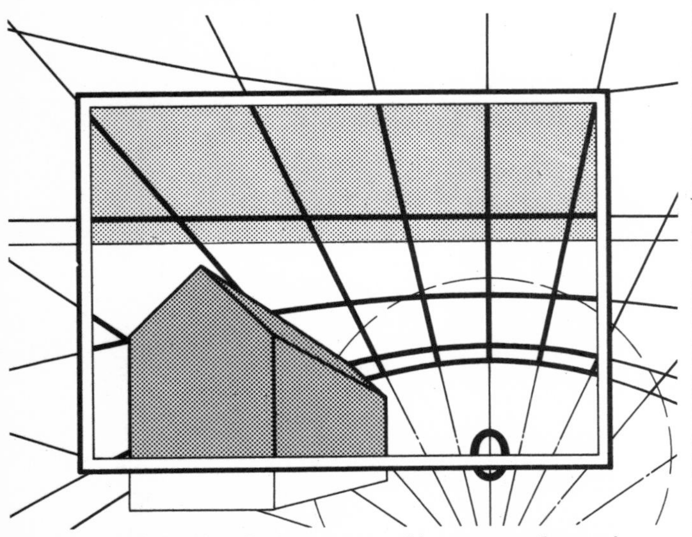

4 *Horizontal shading device can be designed to exclude sunshine at any chosen time and season.*

Table I Storage factors and time lags for buildings of various average densities

Average density of the structure (kg/m³)	Without carpets: Storage factor	Without carpets: Time lag (hours)	With carpets: Storage factor	With carpets: Time lag (hours)
250	0·95	1	0·94	1
500	0·88	1	0·89	1
750	0·83	1½	0·85	1
1000	0·78	2	0·81	1½
1250	0·74	2	0·78	1½
1500	0·71	2	0·76	1½
1750	0·68	2	0·74	1½
2000	0·65	2	0·71	1½
2250	0·62	2	0·70	1½
2500	0·61	2½	0·69	2

Table II Ranges of values of glass properties. Single glazings and sealed double-glazing units are included.

Property	Range of values
Visible transmittance	0·07–0·89
Solar heat reflectance	0·05–0·51
Solar heat absorptance	0·08–0·82
Solar heat transmittance (direct)	0·07–0·85
Shading coefficient	0·18–1·01

Table III The transmittance of sloping glass for radiation direct from the sun at its zenith

Latitude north or south	Angle of window from vertical: 0°	5°	10°	15°
50°	0·70	0·64	0·56	0·43
55°	0·74	0·70	0·64	0·53
60°	0·76	0·73	0·69	0·63

Table IV Shading coefficients for a selection of windows with blinds

Window design	Blind performance	Opaque louver material, Shading coefficient: Short wave	Long* wave	Total	Translucent louver material, Shading coefficient: Short wave	Long* wave	Total
Single glazing without blind							
Clear glass	—	0·92	0·05	0·97	0·92	0·05	0·97
Heat absorbing glass	—	0·51	0·18	0·69	0·51	0·18	0·69
Single glazing with louvers closed							
Clear glass	High	0	0·36	0·36	0·38	0·16	0·54
„ „	Medium	0	0·50	0·50	0·38	0·25	0·63
„ „	Low	0	0·63	0·63	0·38	0·34	0·72
Heat absorbing glass	High	0	0·40	0·40	0·21	0·27	0·48
„ „ „	Medium	0	0·46	0·46	0·21	0·31	0·52
„ „ „	Low	0	0·52	0·52	0·20	0·36	0·56
Single glazing with louvers at 45°							
Clear glass	High	0·10	0·43	0·53	0·34	0·19	0·53
„ „	Medium	0·07	0·55	0·62	0·31	0·31	0·62
„ „	Low	0·05	0·65	0·70	0·28	0·42	0·70
Heat absorbing glass	High	0·05	0·41	0·46	0·17	0·29	0·46
„ „ „	Medium	0·03	0·47	0·50	0·15	0·35	0·50
„ „ „	Low	0·02	0·52	0·54	0·14	0·40	0·54
Double glazing without blind†							
Clear glass	—	0·74	0·10	0·84	0·74	0·10	0·84
Heat absorbing glass	—	0·41	0·14	0·55	0·41	0·14	0·55
Double glazing with louvers between, closed†							
Clear glass	High	0	0·15	0·15	0·32	0·10	0·42
„ „	Medium	0	0·21	0·21	0·31	0·14	0·45
„ „	Low	0	0·26	0·26	0·31	0·18	0·49
Heat absorbing glass	High	0	0·19	0·19	0·17	0·16	0·33
„ „ „	Medium	0	0·21	0·21	0·17	0·17	0·34
„ „ „	Low	0	0·23	0·23	0·17	0·19	0·36
Double glazing with louvers between, at 45°†							
Clear glass	High	0·09	0·19	0·28	0·28	0·12	0·40
„ „	Medium	0·05	0·24	0·29	0·25	0·16	0·41
„ „	Low	0·04	0·27	0·31	0·23	0·20	0·43
Heat absorbing glass	High	0·04	0·20	0·24	0·14	0·16	0·30
„ „ „	Medium	0·03	0·22	0·25	0·13	0·18	0·31
„ „ „	Low	0·02	0·23	0·25	0·11	0·20	0·31
Double glazing with internal louvers, closed†							
Clear glass	High	0	0·40	0·40	0·31	0·22	0·53
„ „	Medium	0	0·50	0·50	0·31	0·29	0·60
„ „	Low	0	0·60	0·60	0·31	0·35	0·66
Heat absorbing glass	High	0	0·32	0·32	0·17	0·22	0·39
„ „ „	Medium	0	0·37	0·37	0·17	0·25	0·42
„ „ „	Low	0	0·42	0·42	0·17	0·28	0·45
Double glazing with internal louvers, at 45°†							
Clear glass	High	0·08	0·44	0·52	0·27	0·25	0·52
„ „	Medium	0·05	0·53	0·58	0·25	0·33	0·58
„ „	Low	0·04	0·60	0·64	0·22	0·42	0·64
Heat absorbing glass	High	0·04	0·32	0·36	0·14	0·22	0·36
„ „ „	Medium	0·03	0·37	0·40	0·13	0·27	0·40
„ „ „	Low	0·02	0·40	0·42	0·11	0·31	0·42

* Includes long wavelength radiation and convected heat.
† In all the double glazing systems the inner glass is clear.

Heat loss

Consider heat loss and get an idea of how the energy is lost

7.15 Heat is lost to the outdoors through the glazing mostly by conduction because radiation from the comparatively cool interior is at such long wavelengths that the glass is virtually opaque to it. Glass has a relatively low coefficient of thermal conductivity, 1·05 W/m K, but it is used in buildings in such thin layers that most of the resistance to heat transfer provided by a window is due to the layers of comparatively still air close to the glass surface. For a single sheet of glass, doubling the thickness from 6 mm to 12 mm increases the overall thermal resistance by only 3 per cent.

Consider modification of this heat loss by the insulating properties of single and double glazing

7.16 To describe and compare the thermal properties of the structural elements of buildings, a unit that includes the resistive effects of all the component layers (materials, air spaces and surface layers) has been defined.

7.17 The thermal transmittance, U, evaluates the rate at which heat is transferred from the air on one side of a structure to the air on the other side for unit area and for unit air temperature difference. Values for a selection of windows are shown in Table V.

7.18 Because of the small contribution from the glass towards the insulating effect of the window it is logical to increase the thermal resistance by using two panes of glass separated by an air space rather than to increase the thickness of the glass. As the air space width is increased the thermal transmittance (U-value) decreases and the insulation improves.

7.19 For air spaces more than about 20 mm wide the thermal transmittance is substantially constant because the reduction in conduction transfer is offset by the increase in convection. Below this 'optimum' separation of the panes different methods of increasing the thermal insulation of double glazing can be used.

Consider higher performance double glazing

7.20 It has been known for some time, with opaque structures, that lining air spaces with aluminium foil can increase the resistance to heat transfer by reducing the radiation component. This principle can be applied to sealed double glazing units: the outer pane can be a heat reflecting glass that has a reflective layer on its inner surface.

7.21 Replacing the air between the panes of glass by gases of different properties can alter the thermal properties of the unit.[28] The thermal conductivity, density and viscosity of the gases will all affect the transmission of heat through them. Intuitively it would be expected that the gas with the lowest thermal conductivity would give the best result but this does not always happen because of the effect of convection.

7.22 The influence of the gas within the space is most marked at narrow pane separations and in conjunction with cavity surfaces of low emissivity.

Consider insulation of window frame

7.23 In many situations, such as where compliance with the building regulations is being discussed, the relevant value of thermal transmittance is that for everything that fills the structural opening, the glazing and the associated frames. The effect of the frame on the U-value of the window can be substantial, as shown in Table VI. In general, wooden frames reduce the overall thermal transmittance while aluminium frames not equipped with thermal barriers increase it.

7.24 Apart from some minor aspects, such as the effect of the height of the air space and the influence of sill design on air movement, the heat loss through a window does not depend upon the shape but only the thermal transmittance, the area, and the difference between the temperature of the air on the two sides of the window.

Table V The thermal transmittances of a selection of windows for the three degrees of exposure defined by the IHVE. The reflective units have an emissivity of 0·1 on one air space surface.

Types of glazing	Thermal sheltered	Transmittance normal	(W/m²K) severe
Single	5·0	5·6	6·7
Double with 6 mm air space	3·2	3·4	3·8
Double with 12 mm air space	2·8	3·0	3·3
Double with 20 mm air space, or more	2·8	2·9	3·2
Reflective double unit, 6 mm air space	2·4	2·5	2·7
Reflective double unit, 12 mm air space	1·7	1·8	1·9

Table VI Thermal transmittance of clear glass and windows

Glass	Frame	Percentage frame area	Average U-value (W/m²K) for exposure Sheltered	Normal	Severe
Single	—	—	5·0	5·6	6·7
	Wood	10	4·6	5·2	6·2
		20	4·3	4·8	5·7
		30	3·9	4·4	5·1
	Aluminium (without thermal barrier)	10	5·0	5·6	6·7
		20	5·0	5·6	6·7
		30	5·0	5·6	6·7
Sealed double glazing units 6 mm air space	—	—	3·2	3·4	3·8
	Wood	10	3·0	3·2	3·5
		20	2·8	3·0	3·3
		30	2·6	2·7	3·0
	Aluminium (without thermal barrier)	10	3·4	3·6	4·1
		20	3·6	3·8	4·4
		30	3·7	4·1	4·7
Sealed double glazing units 12 mm air space	—	—	2·8	3·0	3·3
	Wood	10	2·6	2·8	3·1
		20	2·5	2·6	2·9
		30	2·3	2·4	2·7
	Aluminium (without thermal barrier)	10	3·0	3·3	3·6
		20	3·2	3·5	4·0
		30	3·5	3·8	5·3
Double glazing air space at least 20 mm	—	—	2·8	2·9	3·2
	Wood	10	2·6	2·7	3·0
		20	2·4	2·5	2·8
		30	2·2	2·3	2·6
	Aluminium (without thermal barrier)	10	3·0	3·2	3·6
		20	3·2	3 4	3·9
		30	3·5	3·7	4·2

* These are percentages of the projected areas. Because they take no account of the increased surface that any frame presents to the air, the tabulated U-values for framed windows are somewhat too low.

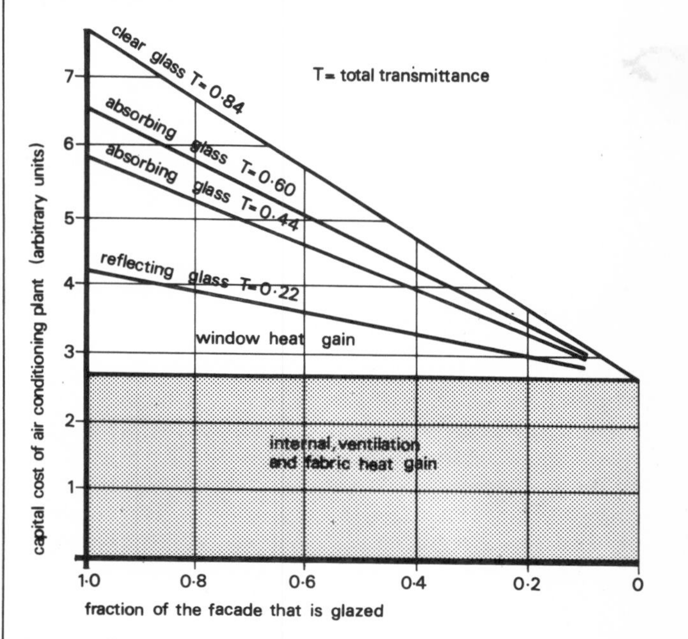

5 *The effect of glass type and window area on the variable capital cost of air conditioning plant.*

Heat loss versus heat gain

Consider energy conservation and get an idea of the balance of heat gain and heat loss.

7.25 Two aspects of window design are important to energy conservation: *solar control* and *thermal insulation.*
Solar control should be designed to make adequate use of the available heat during the heating season but avoid the need to expend energy on the extraction of excess solar heat during the cooling season.

7.26 The capacity of an air conditioning plant, and therefore the capital cost, is determined by the peak values of the solar heat gain to the building. Because the cost of the air conditioning plant in a modern building is likely to be at least a quarter of the whole cost of the building and because up to a quarter of the plant may be needed to handle the solar heat gain through poorly designed windows considerable savings in capital expenditure are possible with careful window design.

7.27 A particular, unexceptional example of the effects of glass type and window area on the variable part of the capital cost of an air conditioning plant is shown in **5**. It does not include the fixed costs, such as installation costs that do not depend upon the size of the plant.

7.28 The running costs of air conditioning depend upon average values of heat gain and not peak values but the general effects of window design on running costs are also, to a first approximation shown by **5**. In this case there are no fixed costs to be added and there may be interactions, such as an increase of internal gains arising from any increase in artificial illumination needed to compensate for reduced window area or glass transmission.

7.29 Over-compensation is to be avoided as is any tendency to provide lighting levels higher than those recommended for the task. When these principles are abandoned, as in some recent buildings, the internal heat gained from the lights can be so great that to keep the air temperature down to the maximum allowed by the present fuel economy regulations requires the expenditure of extra energy.

7.30 Solar heat gain through windows is not always unwelcome and it can make a significant contribution to the overall energy requirement of a building. Burberry[29] has shown that an unobstructed south-facing window gains more heat than it loses during the heating season provided that it is effectively curtained at night. Even when curtains are not closed at night the contribution of solar radiation is considerable as can be seen in **6**, in which the lower of each pair of curves shows the conduction/convection heat loss and the upper the aggregate gain or loss.

7.31 When windows cannot be oriented so as to take advantage of the available solar radiation it is especially important to energy conservation that heat loss through the windows should be kept as low as possible during the heating season.

7.32 The annual expenditure on the heat that is lost through any element of the structure is directly proportional to the area and thermal transmittance (U) of the element, the temperature difference from indoors to outdoors, the duration of the heating season and the cost of fuel. A reduction in any of these will effect a saving and the amount saved can be estimated from diagrams such as that in **7**. The example is worked for a reduction in U-value of 2·6 W/m²K, approximately equivalent to a change from single glazing to double glazing, and for a temperature difference of 12 kelvins, a typical value for general office spaces in central England.

Consider legislation for energy conservation

7.33 In England and Wales, since 31 January 1975, new dwellings must meet improved standards of insulation. The amended building regulations[30] require the average thermal transmittance of the whole of the perimeter walling, including windows and other openings, to be no more than 1·8 W/m²K. Expressed in this form the regulations are not restrictive to design because they allow variations of several of the component factors so that a deficiency in one may be offset by a surplus in another. The extremes of the possible range are exemplified by multiple-dwelling buildings such as blocks of flats and single-dwelling buildings—detached houses.

7.34 A flat that is square on plan and not at a corner of the building satisfies the regulations if 25 per cent of the perimeter of the flat is single glazed. This allows the whole of the one external wall of the flat to be glazed. There is thus little restriction to the acceptable size of windows in blocks of flats and they may be designed to meet some other criterion such as the provision of daylight or an adequate view.

7.35 In detached houses with opaque external walls that are insulated only to the minimum required to meet the legislation ($U = 1 \cdot 0$ W/m²K), the regulations restrict the area of single glazing to 17 per cent of the area of the perimeter, but the design freedom is such that double glazing may be used to increase the window area, **8**. If all the windows are double glazed and the opaque wall is insulated to the practical maximum (say, $U = 0 \cdot 5$ W/m²K) the glazed area may be as much as 56 per cent of the area of the perimeter. Reflective double glazing units are rarely used in private houses but, with a U-value of 1·8 W/m²K which equals the required average, the area glazed with them would be virtually unrestricted.

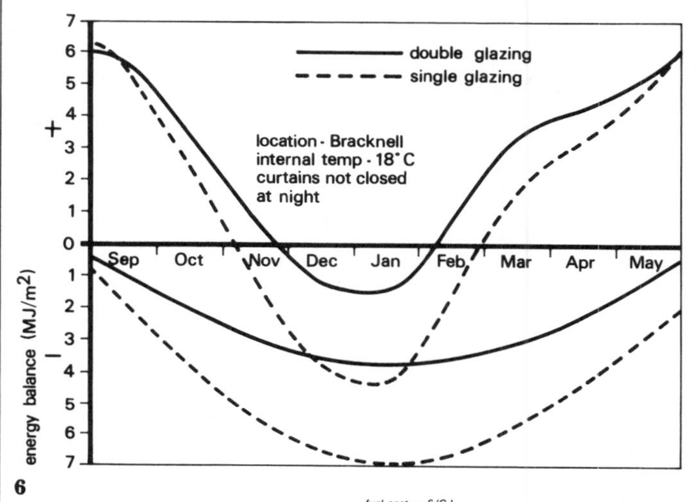

6

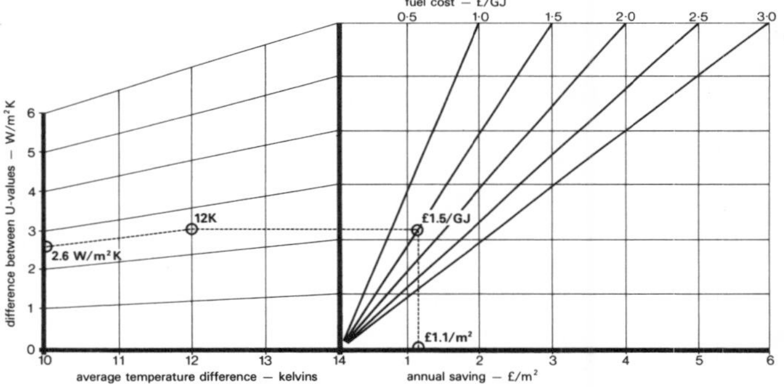

7

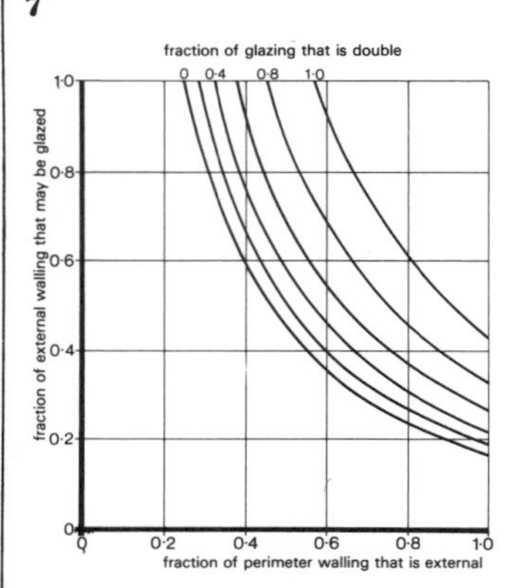

8

6 *Daily energy balance through south, facing glazing.*
7 *The saving due to improving the insulation for a given difference between internal and external air temperatures.*
8 *Permitted areas of various types of glazing. Thermal transmittance of opaque eternal wall,* 1·0 *W/m²K.*

Comfort as a function of heat balance

In considering the windows as a heat filter, comfort depends mostly upon avoiding both 'too hot' and 'too cold'

7.36 Changes in design and methods of construction in the early part of this century exploited new materials and new methods and achieved economies in manpower and in the cost of construction on the site. Many comparatively light-weight structures with large areas of glazing were built at a time when the design expertise was not available to cope with the resulting environmental loads.

7.37 During the same period, standards of comfort acceptable at work, recreation and rest were rising so that the provision of a satisfactory indoor environment was progressively more difficult.

7.38 Moreover, as outdoor noise levels in urban areas have increased, indoor conditions have been adversely affected by large solar heat gains to buildings that were dependent upon natural ventilation and now require the windows to be closed for sound insulation when they should be open for cooling.

Glass selection—avoidance of 'too hot'

7.39 These influences led to a search for better solutions than could be achieved by the expedient of reducing window area and encouraged glass manufacturers throughout the world to produce a wide range of solar control glasses. The properties of a representative selection of the types of glass available are shown in Function 8, Table XII, page 23.

7.40 With careful choice of the glass in relation to the design of the window and the building it is now possible to keep people comfortable in buildings in summer time, with the possible exception of someone who is obliged to sit in direct sunshine for whom an opaque shading device may be necessary.

Glass selection—avoidance of 'too cold'

7.41 The sensations of cold experienced near a single-glazed window when the external air temperature is low are mostly due to loss of body heat by radiation to the cold surface of the glass. It is common to use the term 'downdraught' for the whole of the cold sensation but recent research[31] has shown that radiation is the predominant factor and that the effect of true downdraughts is limited to a region very close to the glass.

7.42 The discomfort can be reduced by installing double glazing. The surface temperature of the inner glass of double glazing is higher than that of single glazing under equivalent conditions and this raises the local mean radiant temperature and reduces the area of the room where discomfort due to radiation and downdraught can be felt. **9** shows the order of improvement.

Consider condensation

7.43 A second consequence of the higher glass surface temperature on double glazing is a reduction in the condensation that obstructs the view through windows in some circumstances and can run off to collect in pools on window sills, damage woodwork and spoil decorations.

The avoidance of condensation

7.44 Four quantities govern the formation of condensation: the internal air temperature, the external air temperature, the relative humidity and the thermal transmittance of the glazing. When three of these factors are known the fourth can be derived from the chart in **10**.

7.45 For example, when the internal air temperature is 20°C and the relative humidity is 60 per cent, condensation will not form on double glazing, U-value 3·0 W/m²K, until the external temperature falls below —3°C. Under the same internal conditions condensation would form on single glazing, U-value 5·6 W/m²K, when the external temperature fell below 8°C.

7.46 If the relative humidity of a space is 100 per cent and the internal temperature is higher than the external temperature, no amount of thermal insulation can prevent condensation forming on the walls and windows. Under these extreme conditions the moisture should be extracted as close to the source as possible to prevent it dispersing through the building.

7.47 Condensation within the air space of sealed double-glazing units is prevented by ensuring during manufacture that there is not enough water vapour present for condensation to occur under practical conditions. Double windows (other than sealed units) should have the air space ventilated to the outdoors to minimise the chance of air space condensation.

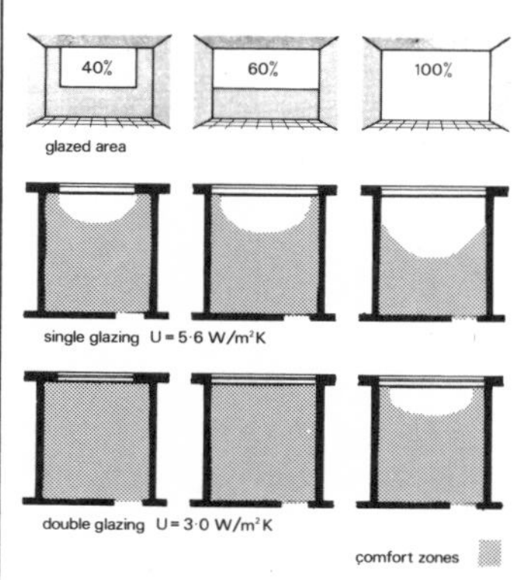

9 *Typical zones of comfort for various window designs.*

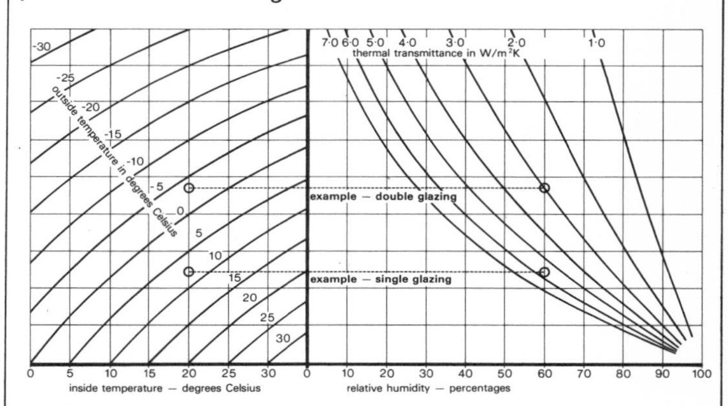

10 *Condensation prediction chart.*

Part I Window Design

1 Window glass design guide

Function 8: glass selection

Basic types of flat glass

Clear float or polished plate glass

8.01 Transparent. Manufactured either by the float process or by the mechanical grinding and polishing plate process. Both types have flat and parallel surfaces, free from the distortion associated with sheet glass. See Table I.

8.02 Clear polished plate glass can be supplied, for special purposes, in thicknesses up to 38 mm and enquiries should be made to the manufacturers. Qualities are:

- GG—glazing quality—for general glazing;
- SG—selected quality—for better class work and mirrors.

Body tinted float or polished plate glass

8.03 Solar control. Transparent. Properties given in Table II.

Surface modified tinted float glass

8.04 Solar control. Transparent. Properties given in Table III.

Clear sheet glass

8.05 Transparent. Manufactured by the flat drawn process, sheet glass has natural fire finished surfaces. Some degree of line distortion is inevitable due to the method of manufacture. The range of thicknesses and the normal maximum manufacturing sizes are given in Table IV.

8.06 Three qualities are available:

- OQ: ordinary glazing quality; for general purposes in factories, housing estates, etc;
- SQ: selected glazing quality; for glazing in buildings requiring a better quality of glass;
- SSQ: special selected quality; for high grade work such as pictures, cabinet work etc.

8.07 An inferior quality is available for horticultural purposes in 3 mm thickness in a range of standard sizes, also in 3 mm and 4 mm in Dutch light sizes to maxima of 1778 × 730 mm and 1000 × 800 mm respectively. 2 mm sheet glass (1·8–2·2 mm) is also available, but is not recommended for glazing, except in small sizes up to 0·1 m².

Body tinted sheet glass

8.08 Solar control. Transparent. Properties given in Table V.

Coloured sheet glass

8.09 Not solar control. Transparent. Available in a variety of colours, the colour may be flashed in a thin layer on to clear sheet, when it is known as 'flashed colour', eg flashed blue, or the whole thickness of the glass may be tinted, when it is called 'pot colour', eg pot blue. Of one quality only and inferior to that of ordinary glazing quality sheet glass. Table VI.

Rough cast glass

8.10 Translucent. Manufactured by the rolling process, this glass has one surface textured with a distinctive nobbly pattern. Properties given in Table VII.

Table I Clear float or polished plate glass

Nominal thickness mm	Thickness range mm	Approx mass kg/m²	Normal maximum size mm
3	2·8–3·2	7·5	2140 × 1220
4	3·8–4·2	10·0	2400 × 1300
5	4·8–5·2	12·5	2600 × 1750
6	5·8–6·2	15·0	4550 × 2500
10	9·7–10·3	25·0	6250 × 3250
12	11·7–12·3	30·0	
15	14·5–15·5	37·5	5000 × 2900
19	18·0–20·0	47·5	4400 × 2800
25	24·0–26·0	63·5	

Table II Body tinted float or polished plate glass

	Nominal thickness mm	Thickness range mm	Approx mass kg/m²	Normal maximum size mm
Green	5	4·8–5·2	12·5	4550 × 2500
	6	5·8–6·2	15·0	
Grey and Bronze	4	3·8–4·2	10·0	2400 × 1200
	6	5·8–6·2	15·0	4500 × 2500
	10	9·7–10·3	25·0	
	12	11·7–12·3	30·0	

Table III Surface modified tinted float glass

Nominal thickness mm	Thickness range mm	Approx mass kg/m²	Normal maximum size mm
6	5·8–6·2	15·0	4550 × 2700
10	9·7–10·3	25·0	
12	11·7–12·3	30·0	

Table IV Clear sheet glass

Nominal thickness mm	Thickness range mm	Approx mass kg/m²	Normal maximum size mm
3	2·8–3·2	7·5	2030 × 1220
4	3·7–4·3	10·0	
5	4·7–5·3	12·5	2350 × 2100
6	5·7–6·3	15·0	

Table V Body tinted sheet glass

	Nominal thickness mm	Thickness range mm	Approx mass kg/m²	Normal maximum size mm
Green	5	4·5–5·1	12·0	3300 × 2650
	6	5·2–5·8	14·0	3560 × 2650
Grey	3	3·1–3·4	8·0	2180 × 1800
	5	4·5–5·1	12·0 or	3300 × 2550 3100 × 2650
	6	5·2–5·8	14·0	3300 × 2650
	7	6·2–6·8	16·0	3600 × 2650
Bronze	3	3·1–3·4	8·0	2184 × 1778
	5	4·5–5·1	12·0	3304 × 2642
	6	5·2–5·8	14·0	3304 × 2642
	7	6·2–6·8	16·0	3556 × 2642

Table VI Coloured sheet glass

	Thickness range mm	Approx mass kg/m²	Normal maximum size mm
Flashed colour only	1·5–2·0	4·5	800 × 600
Flashed and pot colour	1·8–2·5	5·5	1100 × 800
	2·5–3·5	7·5	

Table VII Rough cast glass

Nominal thickness mm	Thickness range mm	Approx mass kg/m²	Normal maximum size mm
5	4·7–5·3	12·5	3700 × 1280
6	5·7–6·3	15·0	
10	9·7–10·3	25·0	

Thick rough cast glass

8.11 Translucent. Manufactured by the rolling process in thicknesses greater than 10 mm and generally the glass is ground and polished to give polished plate glass. Surface textures vary according to manufacture.

Body tinted rough cast glass

8.12 Translucent. Manufactured in similar manner to rough cast. Different colours may be available from time to time, but the colour most usually available is blue/green, in 3 mm thickness up to 2540 × 1220 mm and 5 and 6 mm up to 3040 × 1220 mm.

Patterned glass

8.13 Translucent. Rolled glass, one surface of which has a pattern or design texture, generally of a decorative nature. Usually the deeper the pattern the greater the obscuration and diffusion although this is not invariable. Available in 3 and/or 5 mm according to pattern. Normal maximum size is 2140 × 1280 mm in 3 mm and 2140 × 1320 mm in 5 mm thickness. The nominal thickness is taken from the base of the pattern, so that a glass with a heavy pattern will have a greater overall thickness than a glass with a shallow pattern. Some patterned glasses are also available in a range of colours and tints.

Polished wired glass

8.14 Transparent. Rolled glass with wire completely embedded in it and having ground and polished surfaces. Available with two types of wire mesh: georgian (13 mm square) and diamond (20 mm diamond). Thickness is 6 mm only (tolerance 5·5–7·1 mm) and normal maximum size is 3300 × 1830 mm.

Rough cast wired glass

8.15 Translucent. Rolled glass with wire completely embedded in it. The surfaces are similar to those of rough cast glass. The nominal thickness is 6 mm. Normal maximum size is 3700 × 1840 mm. Only one wire mesh type available namely georgian (13 mm square mesh).

Opal glasses

8.16 Translucent/opaque. Opal and opalescent glasses are glasses varying from a faint milkiness in the case of flashed opal to virtual opacity, pot opal. They may be white, coloured or variegated, the degree of opacity dependent on the concentration of minute crystals and the thickness of the glass. Table VIII.

Table VIII Opal glasses

	Thickness range mm	Approx mass kg/m²	Normal maximum size mm
Flashed opal	2·2–2·6	6·0	1800 × 1200
	2·7–3·3	7·5	2000 × 1200
	3·5–4·0	9·5	
	4·0–5·0	11·5	2400 × 1200
	5·0–6·0	14·0	
Pot opal	1·8–2·5	5·5	1100 × 800
	2·5–3·5	7·5	
	3·5–4·2	9·5	

Products processed from basic types

Toughened glass

8.17 Products derived from subjecting annealed glass to a process of heating and rapid cooling which produces high compression in the surface and compensating tension in the centre. Toughened glass for this reason cannot be cut and should not be drilled, surface or edge worked. The treatment endows the glass with greatly increased resistance to impact, loading and thermal shock. When broken, toughened glass dices into comparatively harmless pieces.

8.18 The length to breadth ratio should generally not be greater than 7:1 but in doubtful cases enquiries should be submitted and processor's literature should also be consulted for information regarding edge working, surface working, drilling, dimensional tolerances, shapes, deviation from flatness, etc.

8.19 As rectification to size is not possible after toughening extra care should always be taken to order the correct size required for glazing.

8.20 The toughening process may be used simultaneously to fire in coloured ceramic enamels. The maximum sizes given below are the ones generally available. However, sizes greater than those listed may be available according to the toughening facilities of the processors. Thickness tolerances are the same as for the basic glass types. See Tables IX and X.

Table IX Sizes of toughened clear float and polished plate glass

Type	Nominal thickness mm	Maximum sizes mm	Maximum area limitation m²
Clear float and polished plate	4	2000 × 900	
	5	1520 × 915	
	6†	2600 × 1350	
	10†	*3950 × 1520	
	12	3100 × 2410	
	15	*3950 × 1520	5·0
		3100 × 2410	6·7
	19	*3950 × 1520	4·2
		3100 × 2410	5·6

** If one dimension exceeds 1520 mm, the other cannot exceed 3100 mm.*
† Also available with coloured ceramic enamel finish.

Table X Maximum sizes of basic types of toughened glass

Type	Nominal thickness‡ mm	Maximum sizes mm
Body tinted	4	2000 × 900
Float or polished	6	2600 × 1350
Plate glass and	10	2700 × 2000
Surface modified tinted float	12	3100 × 2410
Sheet glass	5	1520 × 915
	6	2600 × 1350
Rough cast	6†	2600 × 1350
	10†	3050 × 1800
Patterned*	6	1520 × 915
	10	2600 × 1350

** Enquiries should be made to processors for available patterns.*
† Also available with coloured ceramic enamel finish.
‡ Subject to availability according to glass type—refer to basic types.

8.21 Toughened glass is made with coloured ceramic pigment fired into one surface for cladding panels and insulating infill panels. The coloured surface results in near opacity for normal spandrel and cladding panels having back-up walls, but where the glass is to be used as free standing panels such as balcony glazing and partitions, the use should be made known at the time of ordering to ensure that the product is opaque by manufacture.

8.22 The coloured glass being toughened is not affected thermally outside its safe stress capabilities by solar radiation. It can therefore be used with confidence for cladding in front of any backing materials and insulating infill panels are available. Such panels have a layer of an insulating material bonded to the ceramic surface and a backing of aluminium foil, plasterboard, hardboard, plastics sheet or other suitable materials applied according to building interior requirements.

Laminated glass

8.23 Produced by bonding together two or more panes of glass with polyvinyl butyral ('vinal') interlayer(s). When broken the glass does not separate from the interlayer to any significant extent.

8.24 Thicknesses generally available are from 4·4 mm to 9·1 mm with interlayer 0·38 mm to 1·1 mm and in sizes up to 2440 × 1600 mm.

8.25 Bandit or antibandit glass has an interlayer at least 1·14 mm thickness (greater according to application) and is used for special glazing purposes such as safety against impact and resistance to smash and grab raids.

8.26 Bullet resistant glass is laminated glass, made up of three or more panes of glass with interlayers, with an overall minimum thickness of 22 mm and can be supplied up to 78 mm thick, the thickness being related to the ability of the product to resist a specified level of attack. BS 5051: Part 1: 1973 deals with bullet-resistant glazing for interior use and specified requirements for attack resistance. Another part of this specification (at present being compiled) will deal with exterior use. Laminated glass processors should be consulted when there is any doubt regarding correct specification.
8.27 Solar control laminated glass has either a thin transparent metallic deposition overall on one glass surface adjacent to the interlayer or a tinted interlayer. The former controls primarily by heat reflection, the latter by heat absorption. See Table XI.

Table XI Sizes of reflecting laminated glass

	Nominal thickness mm	Thickness range mm	Maximum size mm
Reflecting laminated glass	6	5·6–6·8	1980 × 1220
	8·5	7·8–9·0	2130 × 1220
	10·5	9·6–10·8	3650 × 2430

8.28 Heat absorbing laminated. The thicknesses available range from 4·4 mm to 12·0 mm in sizes up to 2440 × 1525 mm, with tinted interlayers from 0·38 to 1·5 mm thick.
8.29 Alarm laminated glass incorporates fine wires within the interlayer and is installed in such a way that if a wire is broken an alarm is activated.

Insulating glass units

8.30 Glazing units consisting of two or more panes of glass for the purpose of heat control/insulation. The panes are hermetically sealed at their periphery so as to contain dehydrated air between them. Air space widths are from 3 mm to 12·5 mm and generally any of the range of flat glasses, annealed or toughened, may be used in fabrication.

Surface treated

8.31 Manufacturers' literature should be consulted for detailed information regarding acceptable glass combinations, sizes, shapes, etc. For decorative and/or functional effects, the different methods of surface treatments are described below:
- *Obscuring* involves treating the whole or part of the glass surface so that vision through is obscured.
- *Brilliant cutting* is a decorative process giving designs by various types of cut which are subsequently smoothed and polished. The cuts according to the resultant shape have descriptive names eg V cut and panel cut.
- *Engraving* is achieved by cutting the glass surface with a small revolving wheel, with or without abrasive.
- *Enamelling, staining or painting and firing* is the process of coating with a fusible pigment and subsequently firing to give permanence by chemical combination.
- *Stoving* is the process of coating with a pigment at a temperature lower than that used for firing, the resulting finish relying on adhesion to the glass surface.
- *Gilding* is the application of leaf metal to a glass surface and protecting with a coating; used for lettering and decoration.
- *Silvering* involves the deposition of silver onto the glass and giving protection to the silver by coatings or coverings according to the intended use and the degree of protection.
- *Striped silvering* (venetian silvering) is the process which produces alternate bands of silver and clear glass to form what is often referred to as one-way vision glass, but it is also used for partially obscuring or as a decorative finish.
- *Applying metallic film* is a way of giving glass increased heat and light reflecting/absorbing properties and glass treated in this way generally incorporates permanent protection of the film.

Edge treated and bevelled

8.32 The edge finishes for glass range from 'arris' (which is a bevel not greater than 0·15 mm at an angle of about 45° to the glass surface), achieved by grinding, smoothing or polishing, to such highly skilled bevelling as 'scalloped bevel', 'fluted bevel' and 'crossed bevel'. The various types are named and described in BS 952.

Bent

8.33 Bending is the formation during heating of a curved shape for all or part of a pane of glass. Annealed glass after bending must be re-annealed, but glass which is to be toughened and bent generally has the two processes carried out simultaneously. BS 952 gives details of standard curves, including domes.

Typical performances of a representative range of solar control glasses

8.34 When reading Table XII opposite the following definitions and comments apply:
1 *Visible light.* Light having a spectral distribution corresponding to the CIE. (Commission Internationale de l'Eclairage) Standard Illuminant C. This is approximately the same as daylight.
- Transmittance = the fraction of visible light at normal incidence that is transmitted through the glazing.
- Reflectance = the fraction of visible light at normal incidence that is reflected from all the surfaces of the glazing.

2 *Solar radiant heat.* This is approximately the same as the total radiation (ultra violet, visible and infra-red) that is received at sea level directly from the sun at an altitude of 30°.
- Reflectance = the fraction of solar radiant heat at normal incidence that is reflected from all surfaces of the glazing.
- Absorptance = the fraction of solar radiant heat at normal incidence that is absorbed in all the layers of the glazing.
- Direct transmittance = the fraction of solar radiant heat at normal incidence that is transmitted directly through the glazing without change of wavelength.
- Total transmittance = the fraction of solar radiant heat at normal incidence that is transferred through the glazing by all means. It is composed of the direct transmittance and an appropriate fraction of the absorptance.

3 *Shading coefficient.* A number used to compare the solar radiant heat admission properties of different glazing systems. It is calculated by dividing the appropriate transmittance by 0·87 which is the total transmittance of a notional clear single glazing between 3 and 4 mm thick.
- Short wave shading coefficient = the direct transmittance divided by 0·87.
- Long wave shading coefficient = the fraction of the absorptance that contributes to the total transmittance divided by 0·87.
- Total shading coefficient = the total transmittance divided by 0·87.

4 *Thermal transmittance.* The rate of transfer of heat through unit area of the glazing for unit difference of air temperature on the two sides. For the two exposed surfaces, the values of surface resistance quoted in the *IHVE Guide*, Book A, 1970 for normal exposure are used.
5 *Mean sound insulation.* The average of the sound reduction index over the frequency range from 100 to 3150 Hz.
In all cases, except where specified, the thickness of glass is 6 mm or equivalent and the air space is 12 mm.
8.35 The total solar radiant heat transmittance is derived from the sum of the direct heat transmittance and the amount of absorbed energy which is admitted, this amount being determined on recommendations of The Institution of Heating and Ventilating Engineers in the *IHVE Guide*, Book A, 1970, assuming normal exposure conditions (2 m/s wind speed) as applicable to vertical (wall) surfaces.

Table XII Typical performances of a representative range of solar control glasses available in the United Kingdom

	Visible light		Solar radiant heat				Shading coefficient			Thermal trans-mittance W/m²K	Mean sound insulation dB
Single glazing	Trans	Refl	Reflect-ance	Absorpt-ance	Direct trans-mittance	Total trans-mittance	Short wave	Long wave	Total		
Clear glass	0·87	—	0·07	0·13	0·80	0·84	0·92	0·05	0·97	5·6	27
Heat absorbing											
Monolithic											
Green	0·75	—	0·06	0·49	0·45	0·60	0·52	0·17	0·69	5·6	27
Bronze	0·50	—	0·05	0·51	0·44	0·60	0·51	0·18	0·69	5·6	27
	0·20	—	0·08	0·72	0·20	0·42	0·23	0·26	0·49	5·6	27
	0·14	—	0·10	0·76	0·14	0·38	0·16	0·27	0·43	5·6	27
	0·08	—	0·19	0·72	0·09	0·31	0·10	0·26	0·36	5·6	27
Grey	0·40	—	0·05	0·51	0·44	0·60	0·51	0·18	0·69	5·6	27
	0·20	—	0·12	0·64	0·24	0·44	0·28	0·22	0·50	5·6	27
	0·14	—	0·11	0·72	0·17	0·39	0·20	0·25	0·45	5·6	27
	0·08	—	0·17	0·72	0·11	0·33	0·13	0·25	0·38	5·6	27
SM bronze*	0·50	—	0·10	0·34	0·56	0·66	0·64	0·12	0·76	5·6	27
Laminated											
Green	0·73	—	0·07	0·26	0·67	0·75	0·77	0·09	0·86	5·6	31
Light brown	0·55	—	0·06	0·39	0·55	0·67	0·63	0·14	0·77	5·6	31
Medium brown	0·28	—	0·06	0·61	0·33	0·52	0·38	0·22	0·60	5·6	31
Dark brown	0·09	—	0·05	0·79	0·16	0·40	0·18	0·28	0·46	5·6	31
Heat reflecting											
Monolithic											
Clear	0·61	0·35	0·30	0·09	0·61	0·64	0·70	0·04	0·74	5·6	27
	0·46	0·39	0·32	0·15	0·53	0·58	0·61	0·06	0·67	5·6	27
	0·39	0·35	0·30	0·25	0·45	0·53	0·52	0·09	0·61	5·6	27
Light topaze	0·48	0·38	0·28	0·15	0·57	0·62	0·66	0·05	0·71	5·6	27
Green	0·50	0·35	0·29	0·40	0·31	0·43	0·36	0·14	0·50	5·6	27
	0·38	0·39	0·31	0·43	0·26	0·39	0·30	0·15	0·45	5·6	27
Bronze	0·33	0·34	0·29	0·38	0·33	0·45	0·39	0·12	0·51	5·6	27
	0·21	0·35	0·30	0·44	0·26	0·40	0·30	0·16	0·46	5·6	27
Grey	0·30	0·34	0·29	0·35	0·36	0·46	0·41	0·12	0·53	5·6	27
	0·18	0·35	0·30	0·47	0·23	0·38	0·26	0·18	0·44	5·6	27
Silver	0·20	0·27	0·24	0·56	0·20	0·37	0·23	0·20	0·43	5·6	27
	0·14	0·33	0·24	0·60	0·16	0·35	0·18	0·22	0·40	5·6	27
	0·08	0·44	0·47	0·44	0·09	0·23	0·10	0·16	0·26	5·6	27
Gold	0·20	0·24	0·23	0·56	0·21	0·38	0·24	0·20	0·44	5·6	27
	0·14	0·26	0·30	0·56	0·14	0·31	0·16	0·20	0·36	5·6	27
	0·08	0·28	0·35	0·56	0·09	0·26	0·10	0·20	0·30	5·6	27
Laminated											
Bronze	0·17	0·22	0·31	0·60	0·09	0·28	0·10	0·22	0·32	5·6	31
Gold	0·30	0·26	0·42	0·35	0·23	0·34	0·26	0·13	0·39	5·6	31
Gold	0·16	0·37	0·48	0·43	0·09	0·22	0·10	0·16	0·26	5·6	31
Heat strengthened											
Clear	0·50	0·34	0·26	0·21	0·53	0·60	0·61	0·07	0·68	5·6	27
Bronze	0·30	0·34	0·26	0·40	0·34	0·46	0·39	0·14	0·53	5·6	27
Grey	0·24	0·34	0·26	0·43	0·31	0·44	0·36	0·15	0·51	5·6	27
Double glazing											
Clear glass	0·76	—	0·12	0·24	0·64	0·73	0·74	0·10	0·84	3·0	29
Heat absorbing											
Monolithic											
Green	0·66	—	0·07	0·57	0·36	0·48	0·41	0·14	0·55	3·0	29
Bronze	0·44	—	0·06	0·59	0·35	0·48	0·41	0·14	0·55	3·0	29
	0·30	—	0·06	0·69	0·25	0·38	0·29	0·15	0·44	2·8	29
	0·18	—	0·08	0·76	0·16	0·30	0·18	0·16	0·34	2·8	29
	0·07	—	0·19	0·74	0·07	0·20	0·08	0·15	0·23	2·5	29
Grey	0·37	—	0·06	0·59	0·35	0·48	0·41	0·14	0·55	3·0	29
	0·30	—	0·07	0·64	0·29	0·42	0·33	0·15	0·48	2·8	29
	0·18	—	0·13	0·68	0·19	0·32	0·22	0·15	0·37	2·8	29
	0·07	—	0·17	0·74	0·09	0·22	0·10	0·15	0·25	2·5	29
SM bronze*	0·43	—	0·12	0·43	0·45	0·56	0·52	0·12	0·64	3·0	29
Laminated											
Green	0·64	—	0·10	0·36	0·54	0·64	0·62	0·11	0·73	3·0	32
Light brown	0·48	—	0·08	0·48	0·44	0·55	0·51	0·13	0·64	3·0	32
Medium brown	0·24	—	0·06	0·67	0·27	0·40	0·30	0·15	0·45	3·0	32
Dark brown	0·08	—	0·05	0·82	0·13	0·27	0·15	0·16	0·31	3·0	32
Heat reflecting											
Monolithic											
Clear	0·54	0·37	0·33	0·16	0·51	0·57	0·59	0·06	0·65	3·0	29
	0·42	0·41	0·34	0·22	0·44	0·51	0·51	0·07	0·58	3·0	29
	0·35	0·35	0·31	0·33	0·36	0·44	0·41	0·10	0·51	3·0	29
	0·20	0·38	0·45	0·45	0·10	0·17	0·11	0·08	0·19	2·0	29
Light topaze	0·42	—	0·30	0·23	0·47	0·54	0·54	0·08	0·62	3·3	29
Green	0·46	0·37	0·30	0·44	0·26	0·35	0·30	0·10	0·40	3·0	29
	0·35	0·40	0·26	0·47	0·27	0·31	0·25	0 10	0 35	3·0	29
Bronze	0·33	0·29	0·35	0·47	0·18	0·25	0·20	0·08	0·28	1·8	29
	0·30	0·35	0·29	0·43	0·28	0·36	0·32	0·10	0·42	3·0	29
	0·20	0·35	0·30	0·49	0·21	0·30	0·24	0·11	0·35	3·0	29
Grey	0·27	0·35	0·30	0·40	0·30	0·39	0·34	0·10	0·44	3·0	29
	0·19	0·35	0·31	0·51	0·18	0·28	0·21	0·11	0·32	3·0	29
Silver	0·40	0·35	0·40	0·30	0·30	0·36	0·34	0·07	0·41	1·9	29
	0·30	0·42	0·47	0·32	0·21	0·26	0·24	0·06	0·30	1·8	29
	0·18	0·27	0·25	0·59	0·16	0·27	0·19	0·12	0·31	2·5	29
	0·07	0·44	0·48	0·45	0·07	0·16	0·09	0·09	0·18	2·5	29
Gold	0·40	0·33	0·37	0·38	0·25	0·31	0·29	0·07	0·36	1·8	29
	0·36	0·40	0·51	0·31	0·18	0·28	0·20	0·12	0·32	1·8	29
	0·34	0·22	0·40	0·36	0·24	0·30	0·28	0·07	0·35	2·3	29
	0·30	0·42	0·46	0·37	0·17	0·23	0·20	0·06	0·26	1·9	29
	0·26	0·19	0·26	0·60	0·14	0·22	0·16	0·10	0·26	1·8	29
	0·18	0·24	0·23	0·60	0·17	0·28	0·19	0·13	0·32	2·5	29
	0·07	0·28	0·35	0·58	0·07	0·17	0·08	0·12	0·20	2·5	29
Azure	0·47	0·18	0·24	0·51	0·25	0·33	0·28	0·10	0·38	1·8	29
Grey/blue	0·45	0·20	0·39	0·53	0·28	0·34	0·32	0·07	0·39	2·3	29
Purple	0·39	0·15	0·30	0·46	0·24	0·31	0·27	0·09	0·36	2·3	29
Laminated											
Bronze	0·15	0·22	0·31	0·62	0·07	0·18	0·09	0·12	0·21	3·0	32
Gold	0·34	0·26	0·42	0·39	0·19	0·27	0·22	0·09	0·31	3·0	32
	0·14	0·37	0·49	0·44	0·07	0·15	0·09	0·09	0·18	3·0	32
Heat strengthened											
Bronze	0·25	0·33	0·27	0·46	0·27	0·38	0·31	0·11	0·42	3·0	29
Grey	0·21	0·34	0·27	0·49	0·24	0·34	0·28	0·12	0·40	3·0	29

* Surface modified bronze.

Part I Window Design

1 Window glass design guide

Function 9: glazing techniques

Glass assembly

Glass storage and handling on site

9.01 The main contractor should provide adequate dry storage space and protection for all glazing materials but appropriate racking methods should be the glazing contractor's responsibility. On no account should glass be allowed to stand on or lean against materials which may damage the edges or surfaces.
9.02 The storage space should be protected from rain and dust but it must be adequately ventilated to avoid condensation. Glass should not be stacked without interleaving or spacing materials if there is any possibility of moisture or other liquid contamination.
9.03 The glazing contractor will normally replace all glass broken or damaged by his employees but will not be responsible for damage caused by others.
9.04 The traditional white blobs which indicate that an opening has been glazed are not detrimental to ordinary glass providing that whiting is used. On no account must materials containing lime be used as they can cause surface attack of the glass. If labels are attached to the glass the adhesive used must not attack the glass chemically and must be easy to remove without scraping or scratching. On heat absorbing glasses no labelling or other marking should be used, as local heating might cause thermal breakage. Damage to glass after glazing is often caused by attempts made to remove plaster, mortar or concrete spillage. These materials should all be removed when wet because after hardening they are virtually impossible to remove without scratching the glass.

Basic glazing techniques

9.05 This section is adapted from the *Glazing manual* published by the Flat Glass Association, 6 Mount Row, London W1. The manual is under revision and extracts from the 1968 edition are given here.

Rebate and groove dimensions

9.06 Rebates and grooves should be true, rigid, dry and unobstructed before glazing. They must have a minimum depth of 8 mm, except for very small panes up to 0·1 m² where 6 mm is adequate. Large sizes of glass and windows in exposed sites should be at least 9 mm deep.
9.07 The width of rebate is dependent on a number of factors, but some general guidance can be given. If using putty fronting, the width should be a minimum of 11 mm—2 mm back putty, 3 mm glass and 6 mm front putty. When using gaskets in place of glazing compound the width must accommodate two thicknesses of gasket, the glass and the bead. In glazing bent glass it is necessary to have a wider rebate in order to accommodate the slight inaccuracies of the bent glass. The thickness of the distance pieces generally requires to be varied in order to secure the bent glass properly, **1**.

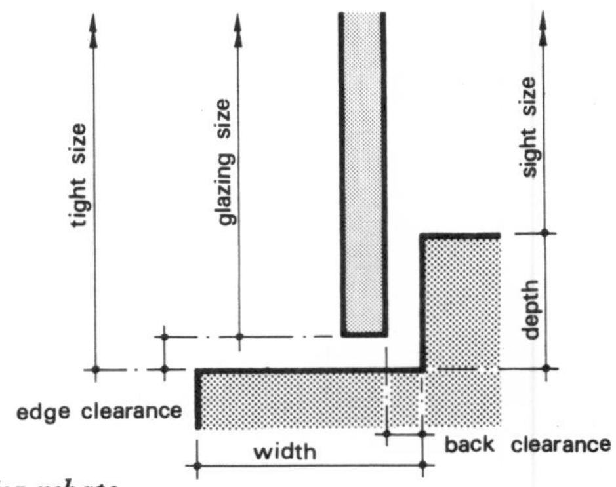

1 *Glazing rebate.*

Preparation of surrounds

Wooden surrounds

9.08 These should be primed to prevent excessive absorption of oil from the linseed oil putty. Frames are frequently supplied already primed, but where this has not been done a satisfactory result can be obtained by applying a coat of primer paint made to BS 2521-24 or an equivalent primer. A shellac varnish or gloss paint should not be used as these completely seal the wood and the setting of the putty will be unduly delayed.
9.09 Absorbent hardwood surrounds which are to be painted subsequently should either be primed with a medium of equal parts of exterior varnish and white spirit and glazed with linseed oil putty or completely sealed with undiluted exterior varnish and glazed with metal casement putty. Water repellent wood preservatives are increasingly being used. Preservatives of high oil and resin content or high solid content show no tendency to interfere with the normal curing of linseed oil putty, and function instead of priming paints.
9.10 Non-absorbent hard woods such as teak do not require priming or sealing and, if putty fronted, metal casement putty should be used, which must be subsequently painted with varnish to prevent deterioration and cracking. If beads are used, non-setting compound is suitable. It should be noted, however, that some absorbent hardwoods are referred to as teak in the building industry and these require to be treated before attempting to glaze. Priming, which is not normally the function of the glazier, is to be carried out by other trades.

Metal surrounds

9.11 Steel frames are usually coated with zinc. Where this has been sprayed on, no further treatment is required. Where hot-dip galvanising has been used, unless the frames have been exposed to natural weathering for some time before glazing, the smooth surface will not satisfactorily accept paint and putty and the surface must be brought to a suitable condition.
9.12 Self-etch two-pack chromate primers based on polyvinylbutyral or zinc chromate primers based on alkyd resins

are recommended. Some manufacturers use chromate passivation on galvanised frames. The use of this material improves the adhesion of oil based glazing compounds.

9.13 If calcium plumbate primers are used they should cause no interference with oil based compounds provided:

1 the plumbate primer is based on a stoved resin finish;

2 air drying finishes based on alkyd resins have had a long (not less than seven days) period in which to attain their full hardness. Due to the difficulty of recognising these differences in plumbate primer composition they should be avoided if possible and the alternatives mentioned above used in preference.

9.14 Mordant solutions should not be used. Aluminium surrounds require coating with zinc chromate priming paint in order to obtain satisfactory adhesion with metal casement putty. Bronze surrounds do not require priming before glazing.

Stone, concrete or brick surrounds

9.15 Rebates or grooves must be sealed with at least two coats of an alkali resistant sealer in order to prevent absorption of the oil and consequential staining. The sealer must also be compatible with the glazing compound. Advice should be obtained from the compound maker. The sealer must be allowed to dry thoroughly before glazing is started. Normally this type of sealer cannot be used at temperatures below freezing point, and storage at higher temperatures is essential.

Glass size and clearances

9.16 When specifying measurements the width should appear as the first dimension. In the case of patterned glass of a line nature, the direction of pattern should be stated, ie vertical or horizontal. In all cases such as patterned glasses, decorated glass, leaded lights or copperlights, where adjacent glasses are required to align, this should be stated.

Location of glass in frame

9.17 The square of glass should be centralised in the surround by resting the bottom edge on setting blocks—of unplasticised PVC or similar material 25 to 75 mm long—the thickness of the blocks being adjusted so as to give equal clearances on all edges. Similar blocks, but known as location blocks, are sometimes used on the top and vertical edges. Details of positioning of both setting and location blocks are shown, **2**. Glasses of area not greater than $0 \cdot 2$ m² may be glazed directly without setting blocks.

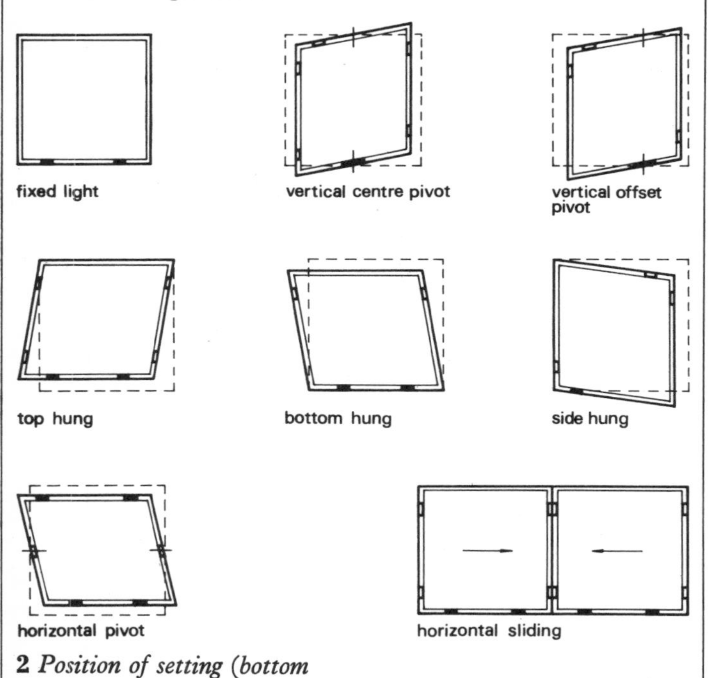

2 *Position of setting (bottom edge) and location blocks.*

Glazing

External glazing with front putty

9.18 Maximum sizes which may be glazed with putty fronting depend on the degree of exposure. As a rough guide, under normal conditions they should not exceed $1 \cdot 5$ m² and for exposed sites this should be reduced to $1 \cdot 0$ m².

9.19 Sufficient putty (linseed oil putty in the case of wooden frames and metal casement putty in the case of metal, sealed hardwood or concrete frames) should be applied to the rebate so that when the glass is pressed into the rebate a bed of compound (known as back putty) will remain between the glass and the rebate. Before actually offering the glass to the surround the setting blocks are pushed into position in the putty in the bottom rebate. The glass is then pushed well back with a duster, care being taken that no voids or spaces are left. The use of a duster is to prevent putty fingermarks on the glass. This action squeezes the surplus putty out of the rebate and should leave a back putty of not less than 2 mm.

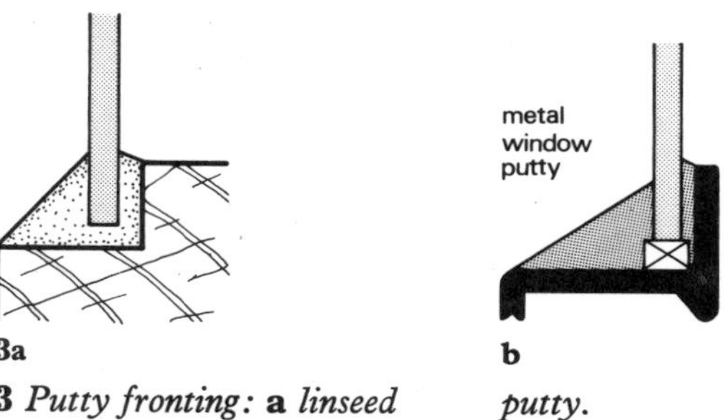

3 *Putty fronting:* **a** *linseed oil putty;* **b** *metal casement putty.*

9.20 The glass should then be secured in wooden windows (with springs tapped into the perimeter not more than $0 \cdot 4$ m apart, and in metal windows by clips or springs fixed into the holes provided), to hold the glass until the putty sets. Further putty should then be applied and finished at an angle from the edge of the frame up to 2 mm below the sight line, thus allowing the painter to paint up to the sight line on the glass and thereby create a seal between the putty and the glass and prevent water from entering. The surplus back putty should be stripped, taking care not to undercut which would cause water lodgment.

9.21 When glazing into grooves the glass should be pressed into glazing compound previously placed into the groove. The spaces between the glass and sides of the groove should then be filled with compound and finished off at an angle. Not undercut, to prevent water lodgment.

9.22 All putties should be finished with a light brushing in order to seal the putty to the glass, **3a, b**. All painting is to be carried out by other trades. Painting is not normally the function of the glazier.

Glazing with beads

9.23 The glazing is done as previously described, with the exception of the front finishing, this being replaced by beads, **4a, b, c, d**.

9.24 A much wider range of compounds may be employed where glazing is with a bead. In addition to linseed oil and metal casement putties, non-setting compounds and one- or two-part polymerising compounds may be used. Preformed strip glazing compounds and gaskets are suitable if compression beads are provided.

9.25 When using linseed oil or metal casement putties, distance pieces are not generally necessary except on sites where displacement is likely to occur before the putty has set, due to wind loading. With non-setting compounds, distance pieces should always be used and they should be of PVC or similar material and be about 25 mm long, depth 3 mm less than the depth of the rebate to permit subsequent cover by compound, and thickness equal to the space between frame and glass and between glass and bead—generally 3 mm. They should be used opposite each other and spaced apart not more than 0·4 m. Where possible they should be located at the fixing points of the beads.

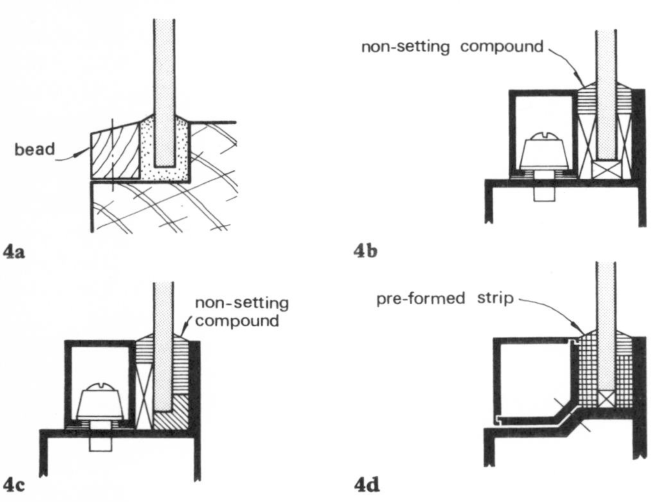

4 *Bead glazing:* **a** *timber bead;* **b** *non-setting compound;* **c** *non-setting compound with sealant 'heel bead';* **d** *strip compound with pressure bead.*

9.26 Where the bead is *outside* it should be bedded with compound against the glass and the rebate, and pressed in to give a depth of compound 3 mm between the glass and the bead. Hollow beads are undesirable unless they can be completely filled with compound. When the beads have been secured in position with an adequate number of pins or screws to prevent flexing or movement, the compound should be finished at an angle so as to prevent water lodgment.
9.27 Where the bead is *inside* it should be bedded against the glass, but it is not necessary for it to be bedded to the rebate.
9.28 In applications where preformed strip materials are used, they will act as a continuous distance piece. Tapes and extrusions in this category can be used as primary seals between smooth joint surfaces, ie glass to metal, but some positive method of compression must be applied to ensure a weathertight seal. They should be of a thickness which, when compressed, is not less than 3 mm and should be of a depth sufficient to protrude above the rebate and so allow trimming without undercutting.
9.29 In the absence of a true compression glazing bead, these tapes can be lightly compressed by selecting oversize materials for the eventual joint size, but to provide a proper seal it is necessary to top point with curing type sealants such as polysulphides. In order to accommodate this pointing, the depth of the strip should be such as to leave adequate space on top to accommodate the sealant, at least 6 mm. However, in situations where the degree of weather exposure is slight, good results have been obtained with less expensive pointing materials such as key grade non-setting glazing compounds. Manufacturers of individual materials should always be consulted before final specification is made, **5**.

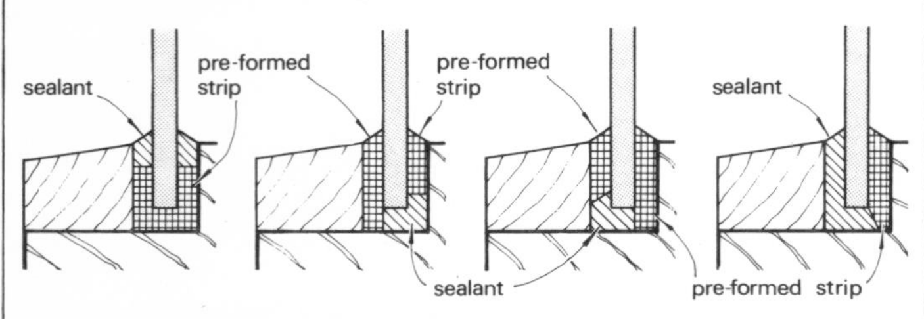

5 *Sealant techniques as illustrated by Tremco Ltd.*

9.30 Bead glazing is also used in conjunction with preformed glazing gaskets where the bead is used to obtain compression on the gasket. The basic principle of gasket glazing is the simple one of providing a permanently elastic and resilient substance which resists pressure from external sources and maintains a durable and weathertight seal between glass, frame and bead surfaces.
9.31 Cleanliness is important during installation and care should be taken to guard against grit or dirt on the surfaces to be sealed. Care must also be taken in the joint design and full account taken of requisite torque loads to be applied to bead screws. Corners can be site mitred and secured with adhesive but in most gaskets this leads to other problems which can be avoided if factory moulded corners are specified.
9.32 It is important that any relative movement between glass and frame is accompanied by distortion of the gasket alone and not by the sliding of the surface forming the weatherseal. Successful application of gasket glazing depends on careful co-operation between gasket manufacturer, window frame designers and the glazing contractor.

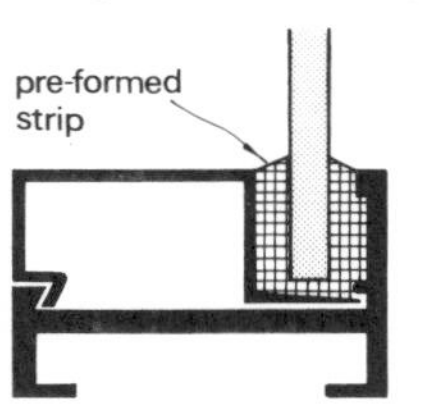

6 *Aluminium spring bead.*

External glazing with structural gaskets
9.33 In this type there are two grooves, one to fit on the surround and the other to accommodate the glass. The compression seal on both components is obtained by inserting a filler strip in a continuous cavity slot on one face. Gasket should have injection moulded corners. Dimensional accuracy is vital, ± 2 mm, if a good weather seal is to be obtained. Gaskets are delivered to the site ready for fitting either direct to the main structure or to subframes which must themselves be weatherproof at all joints. Gasket manufacturers should be given the sight size of the opening—not the glass size. They will then add their own calculated dimension increase to provide slight compression during fixing, **7**.

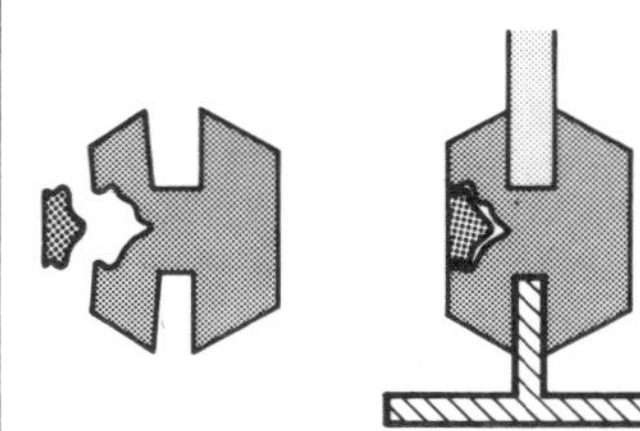

7 *Structural gaskets.*

Double glazing

9.34 *General.* In the case of hermetically sealed units, the design considerations and procedures vary because of the extra thickness, the manufacturing tolerances and the need to protect adequately the edge seal. These are discussed below.
9.35 *Hermetically sealed factory made units.* 1 Ordering of units. Some manufacturers require only the glazing size, but most request that the tight size and sight size of the opening should be specified. Requirements should be checked with the supplier. In taking the measurements, check that any projection, such as that caused by beads or bead retaining grooves and brackets for fastenings, are allowed for. 2 Handling and storage requirements. Units delivered in cases must be unpacked on arrival and if wet should be dried. Units must never be stored flat. They should be stored on edge on racks in dry conditions. Supporting blocks of wood or felt should be used to prevent edge damage. The blocks should be set to form a right angle with the back support so that both panes are supported.

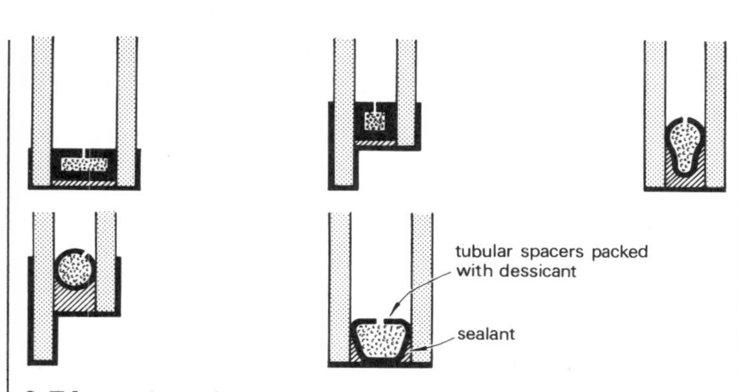

8 *Edge sections of some typical sealed double glazing units.*

9.36 *Design considerations.* It is the responsibility of those manufacturers who require tight and sight sizes to supply units of dimensions which give at least the minimum edge clearance and minimum 9 mm edge cover, see Table 1.

● Clearance and rebate data: Frames, beads and fixings shall be adequate to support and restrain the glazed unit and to withstand wind loadings. The width of the rebate should be sufficient to accommodate the thickness of the unit plus face clearance(s) as required, and plus either the bead or face putty. Face clearances between rebate and glass, and between glass and bead, when used, should be a minimum of 3 mm except in the SMW range of metal windows where the back bedding may be reduced to 2·5 mm, see Table II. When units exceeding 16 mm nominal thickness are to be glazed into opening windows, special consideration may be needed for the design of the frame and for the method of glazing. The frame manufacturer should be consulted.

Table I Design considerations for double glazing

Unit dimensions	Manufacturing tolerances	
	Length and breadth	Thickness
Area m²	mm	mm
Up to 3	+3 −0	+2 −1
3 to 9	+5 −0	+2 −1
9	+6 −0	+2 −1

Table II Face clearance between rebate and glass

Unit dimensions		Minimum edge clearance	Minimum rebate depth*
Area	Thickness		
m²	mm	mm	mm
Up to 3	16	3	12
3 to 9	25	5	16
Over 9	Over 25	Discuss with manufacturer	

* When using sealant capping the depth should be increased by 3 mm.

● Setting blocks: fixed frames should be of sufficient strength to support the unit on two setting blocks positioned at the quarter points. If it is necessary to place the setting blocks closer to the corners in order to avoid excessive deflection they should never be less than 75 mm from the corner. In opening windows the recommendations of the window manufacturers regarding the position should be followed. Setting blocks should be of lead, rigid nylon, sealed hardwood or unplasticised PVC. Generally they should be: *length*, 25-150 mm dependent on the size of the unit; *thickness*, 3 or 5 mm; *width*, 3 mm wider than the unit thickness, except where strip glazing materials are used, when the width should be equal to the width of the unit. The thickness of setting blocks should be such as to locate the unit centrally in the frame.

● Location blocks: are used on the sides and top of opening windows. They should be 25 mm long except in the case of horizontally pivoted windows, where they should be 75-150 mm long to carry the weight of the unit when the window is reversed.

● Distance pieces: should be of plasticised PVC. They should be 25 mm long and should be used in all cases except where strip glazing materials are used. Their thickness should be appropriate to the face clearance allowed. They should be used opposite each other on each side of the unit and positioned not more than 0·4 m apart and adjacent to fixing points of the bead. The first pair of distance pieces should be placed not more than 75 mm from the corner. The height should never be more than 3 mm less than the depth of rebate or bead. In general the position of setting blocks and distance pieces should not coincide. In the case of putty fronting they are used only between unit and rebate. (NB When strip materials are used they must be such that by final compression by the glazier and by wind pressure they cannot be flattened to less than 3 mm.)

● Beads: when wooden beads are used, the dimensions, cross-section and quality should be such that shrinkage, warping, flexing between fixing points and lifting at the corners will not occur. They should be held with screws located not more than 50 mm from each corner and spaced not more than 225 mm apart intermediately. Ideally the screw fixing points should be provided with cups. The fixing points of metal beads held with screws, studs or clips must be located not more than 75 mm from each corner and spaced not more than 225 mm apart intermediately.

● Compounds: care must be taken to ensure that the compounds used are compatible with the edge seal and where more than one compound is used that they are compatible with each other.

● Glazing sizes: all units must be adequate to withstand the exposure conditions.

9.37 *Preparation of surrounds.* All windows which have to be prepared for units must be set in accordance with the frame makers' instructions and any damaged in transit or fixing must be rectified, so as to provide frames which are square, without distortion and with rebates free from obstruction. Concrete frames and wooden windows, whether primed softwood or unprimed hardwood, should be treated with two coats of sealer, including the inside faces of the beads, to prevent absorption of oil from the compound. Where metal frames have been treated with calcium plumbate it is important to ensure that the priming is fully hardened otherwise there may be reaction with the glazing compound. This is done by wiping gently with a cloth moistened with white spirit; any removal of the film will indicate that the primer is not sufficiently dry. Before glazing begins rebates must be dry and free from dust. For checking timber frames for wetness, a sharp object such as a nail file or nail should be pressed on to the bottom surface of the rebate to observe whether water is squeezed out. If glazing with sealants in order to obtain effective adhesion the use of special primers may be necessary on wooden frames. It is necessary to remove protective coatings such as lacquer, wax, priming or paint from all frames. In all cases the manufacturer's guidance should be followed.

9.38 *Site procedure for the glazier.* Check that the units have not been damaged, particularly in the case of units with protective foil. Check each unit for edge clearance in the opening. No attempt must be made to alter the size of the unit where this is not correct, by nipping or other means. Where units have protective foil edging which projects above the sight line, check procedure with the manufacturer because this varies from one to another. Some allow the edging to be trimmed back, others allow the edging to project and stay exposed, while others insist that a deeper rebate be used. Where units are glazed into rebates of minimum depth, part of the edge seal or spacer in some units may project above the sight line. If complete cover is required a frame with a deeper rebate will need to be specified. Generally the compound manufacturers will not accept responsibility unless the edging is covered by at least 3 mm of compound.

Glazing procedures for double glazing units

9.39 There are a considerable number of systems which can be satisfactorily employed.

Bedding with non-setting compound and fronting with metal casement putty or sealant (NB: It is essential that the two materials are compatible.) Table III.

Table III Bedding with non setting compound and fronting with metal casement putty or sealant

	N/m²	m²
Up to	75	1·5
	75 to 100	1·1
	100 to 150	0·8
Over	150	Use beads.

When using this system the above sizes should not be exceeded for wind loadings as tabulated in the left-hand column.

9.40 The sequence of operations is as follows:
1 Check that rebates and beads have been properly prepared (see **9.37**).
2 Apply sufficient non-setting compound to rebate to ensure that when the unit is pressed home there will be a solid bed between the unit and rebate and surplus compound will be squeezed out.
3 Press on to the rebate the appropriate number of distance pieces, of the correct dimensions, at the required positions.
4 Press setting blocks, 3 mm wider than the thickness of the unit, hard on to the rebate at the required positions.
5 Position the unit in the frame and adjust for clearance. Insert location blocks 3 mm wider than the thickness of the unit, as required for single glazed opening windows, and press the unit firmly on to the distance pieces all the way round.
6 Fill completely the spaces between the edge of the unit and the frame, eliminating all voids.
7 Point or trim the compound level and flush with the face of the unit. Fix clips, cleats or sprigs. Clean glass and rebate and then front with metal casement putty or sealant to finish not less than 14 mm in height and to the edge of the rebate. Brush off lightly to seal the putty to the unit.
8 Trim off surplus compound at the back at an angle to shed water.
Maintenance check period—every three years, **9**.

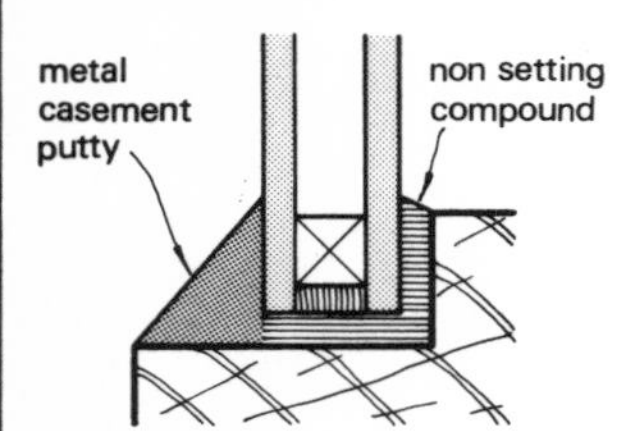

9 *Multiple glazing unit—putty fronting.*

Glazing with non-setting compound and beads

9.41 The sequence of operations follows that of putty fronting, down to 6, and then proceeds as follows:
7 Apply compound to the rebate and unit, press in distance pieces of the correct dimensions in a position to coincide with the distance pieces on the opposite face of the unit. Where channel beads fixed over studs are used, dip the screw thread in compound and fix stud to the frame, then proceed as above.
8 Apply bead and press on to the distance pieces and secure.
9 Point in with compound where necessary and trim off to provide a smooth chamfer.
Maintenance check period—every five years, **10**.

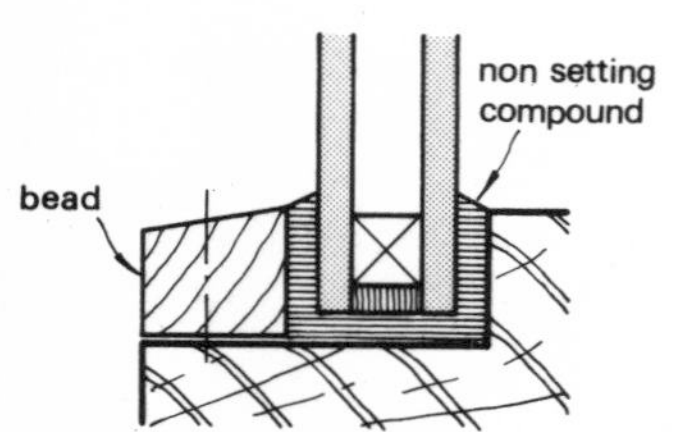

10 *Multiple glazing unit—bead glazing.*

Glazing with preformed strip with beads and capping with non-setting compound

9.42 The sequence of operations is as follows:
1 Check that rebates and beads have been properly prepared (see **9.37**).
2 Apply to the rebate, strip compound of adequate thickness and of a depth to finish at least 3 mm from the top of the rebate when it is contact with the bottom of the rebate. When capping is to be only on the outside, the inner strip should be equal to the depth of the rebate.
3 Place setting blocks, the width of the thickness of the unit, hard against the strip compound at the required positions.
4 Position the unit in the frame and adjust for clearance. For opening windows insert location blocks the width of the thickness of the unit, as required. Press the unit firmly on to the strip compound all the way round.
5 Gun or knife compound into the spaces between the edge of the unit and the frame, completely filling all voids.
6 Apply a thin layer of compound to the rebate.
7 Apply strip compound of adequate dimensions to the unit to finish at least 3 mm from the top of the bead.
8 When only hand pressure is used, round section strip should be used. Press beads firmly home compressing the strip compound and secure the beads to the frame. When pressure is by beads designed to apply pressure mechanically, apply beads and fix, applying the design pressure in the manner provided for by the frame manufacturer.
Many spring-on types of bead do not permit this method of glazing.
9 Apply compound to fill completely the spaces between rebate, bead, and unit and trim off to provide a smooth chamfer.
Maintenance check period—every five years.

Glazing with preformed strip with beads and capping with sealant

9.43 Proceed exactly as in the previous system down to 8, except that the strip when compressed should finish 6 mm from the top of the rebate and bead because greater depth is needed when using sealant capping. Then proceed as follows:
9 Clean the bead, the frame and the glass at the top of rebate with degreasing solvent and finish a further cleaning with a clean dry cloth.
10 Prime as required by the compound manufacturer.
11 Apply sealant with gun, and trim off to provide a smooth chamfer.
Maintenance check period—every ten years.

Glazing entirely with sealant and beads

9.44 This system, although costly, can offer better protection when the unit is to be subjected to severe vibration and shock, such as may arise in sliding doors and also where access for maintenance is expensive.

9.45 The sequence of operations is exactly the same as in **9.41** except that before starting to apply the sealant, 2, the rebate, bead and glass must be thoroughly cleaned with degreasing solvent and priming applied as laid down by the compound manufacturer, otherwise proper adhesion of compound will not be achieved. Where it is necessary to remove from metal frames any protective coatings such as lacquer, wax, priming or paint, the frame and compound manufacturer's instructions must be followed.
Maintenance check period—every twelve years.

Glazing with insert frames

9.46 The system involves metal sections of channel formation with mitred corners which are filled with compound and applied to the unit. When joined at the corners they form a completely framed unit which is bedded and fixed with screws to the metal window. Maintenance check period—this is dependent on which of the various compounds is employed, and the appropriate maintenance period applies.

Compression glazing with gaskets

9.47 It is strongly recommended that when this system is to be used, the unit maker, frame maker, gasket supplier and glazier should be brought together for consultation at the design stage.

9.48 Compression gasket glazing means that the glazing unit in a window frame is held under pressure in position by a compressed resilient section extending along or enclosing the edges of the glass. This system avoids the necessity for providing distance pieces and may avoid the necessity for setting blocks. The gasket must be suitably resilient in order to accommodate the tolerances on the dimensions of the frame, glass, and pressure members, and to hold the glass firmly, thereby ensuring a weathertight seal.

9.49 *Two systems* can be described as gasket glazing: *Type* 1 consists of two strips of resilient material based on polychloroprene (neoprene) or butyl rubber, or some other material of comparable durability and resilience, one on each side of the unit. These strips form a continuous distance piece and when under compression provide a seal. Setting blocks are required. The procedure for glazing is similar to that described in **9.41** and **9.42** above.

9.50 *Type* 2 consists of a special section enclosing the edges of the unit, used to combine the functions of location and sealing without the use of distance pieces and setting blocks. This form of gasket is compressed into contact with the glass and frame and relies on the retention of stress set up during installation for the permanence of the seal. The gasket must also be capable of accommodating the agreed dimensional tolerances of the glass, bead and frame. Due to the fact that this type of glazing relies completely on the compression and resilience of the gasket to provide a permanent seal, materials subject to substantial cold flow must not be used. Since the gasket is not protected by any external application of glazing compound it must have very good weather resistance and satisfactory results have only been obtained so far with good quality *neoprene* or vulcanised butyl rubber-based materials.

9.51 These sealing gaskets can take a number of forms, but the two most commonly used are: the *channel* gasket, which is designed to be compressed by the glazing bead or frame; and the *zipper* gasket, which takes the form of an extruded 'H' section, one groove to accommodate the unit and the other the frame or surround, and having an additional groove on the side into which a zipper strip is inserted to develop the compression seal.

9.52 In both cases, moulded corners, preferably joined by vulcanisation to the extruded section under controlled conditions, are strongly recommended. The gaskets should be supplied as factory fabrications suitable for the glazing size required. The beads and frames must be sufficiently rigid to avoid bowing between fixed points under the pressure applied.

9.53 Some unit manufacturers may require provision to be made for the drainage from the gasket of moisture which may penetrate.

9.54 Gasket rubbers may become stiff and difficult to install when cold. Gentle warming before use is recommended. If a lubricant is used, this must be approved by the manufacturers. Only smooth tools should be used for the insertion of the gaskets and zipper strips.

9.55 The recommendations of the manufacturers should be followed in carrying out the actual insertion of the gasket and the glazing of the unit. NB: the reader is referred to the Insulation Glazing Association, 6 Mount Row, London W1 for their pamphlet which gives specific details of what to look for and remedial treatment necessary.

Patent glazing and dry glazing systems

9.56 Systems which are designed to drain water away and prevent it remaining in continuous contact with the edge of the unit may be acceptable, but in all cases the unit maker should be consulted and his recommendations followed.

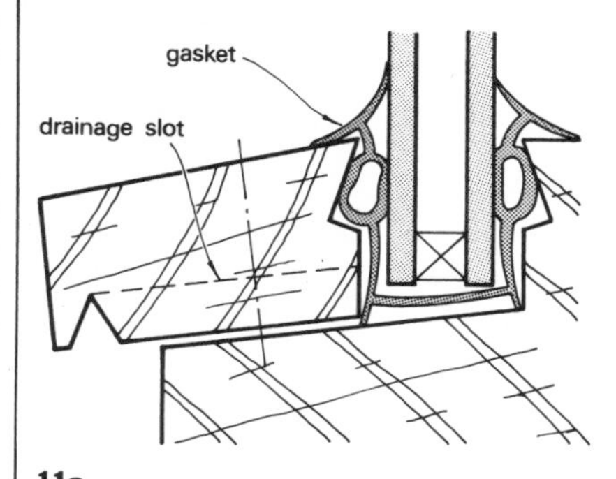

11a

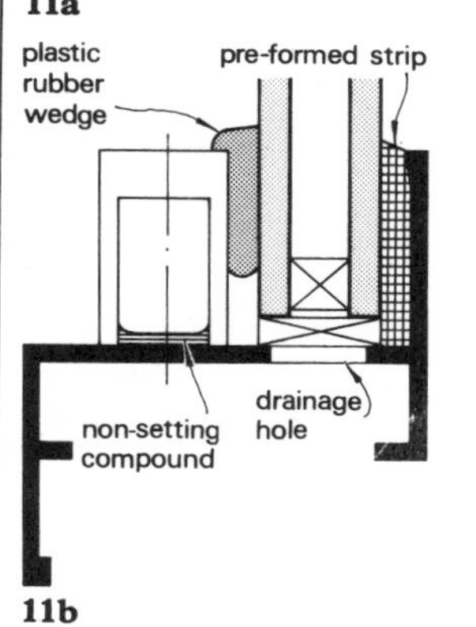

11b

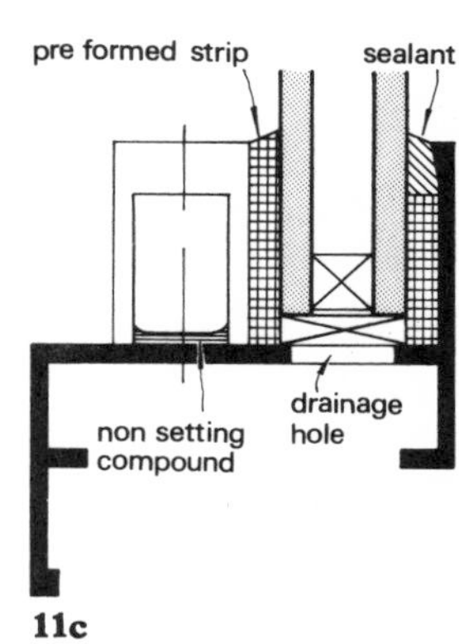

11c

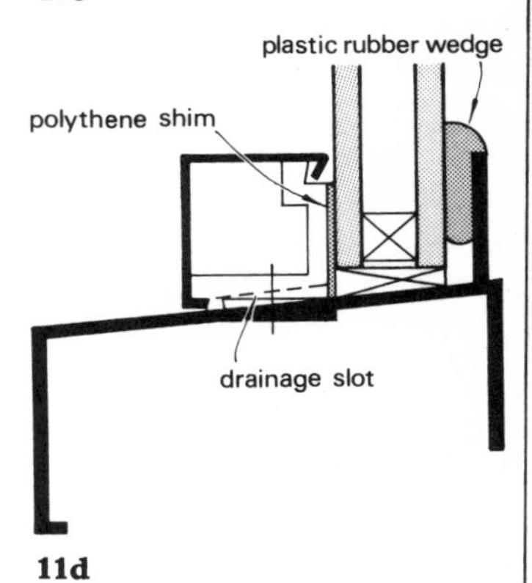

11d

11 *Glazing joints for sealed double glazing units:*
a *Drained joint using gasket.*
b, c *Drained joint examples.*
d *Experimental drained joint designed by BRS.*

Special applications

9.57 Where unusual conditions prevail such as high humidity, chemical fumes, eg in swimming baths, laundries, chemical factories, high sterility units in hospitals and food factories, glazing with sealants (systems described in **9.43** and **9.44**) will generally give satisfactory results. The sealant may have to be specially formulated and the compound maker should be consulted in all cases.

9.58 *Stepped units.* Stepped units can be used where the depth or width of the rebate is inadequate for the glazing of standard units, **12**.

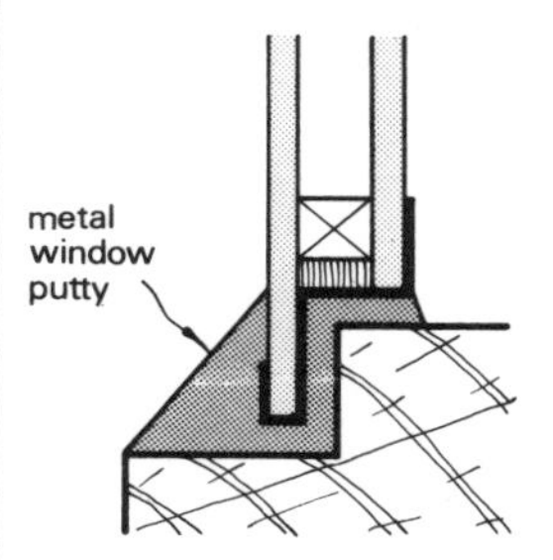

12 *Stepped unit glazing.*

References

1 *Office design: a study of environment* P. Manning (ed). Pilkington Research Unit, Department of Building Science, University of Liverpool. 1965.
2 *Heat reflecting/low light transmission glasses in office building* P. G. T. Owens (ed). AJ 8.5.74.
3 Ludlow, A. M. *The view from the window* To be published by Glass in Building.
4 Markus, T. A. *The significance of sunshine and view for office workers* CIE Intersessional Conference on Sunlight in Buildings, Newcastle upon Tyne, 1965. Proceedings published by Bouwcentrum, Rotterdam, 1967.
5 Keighley, E. C. 'Visual requirements and reduced fenestration in offices' *Building Science* vol 8, no 4, December 1973. Reprinted as Building Research Establishment Current Paper 41/74.
6 DOE Welsh Office *Sunlight and daylight* London. HMSO. 1971.
7 *The flow of light in lighting design* Environmental Advisory Service Report. Pilkington Brothers Limited. 1974.
8 The Illuminating Engineering Society *Daytime lighting in buildings* IES Technical Report no 4 Second edition. London. 1972.
9 British Standards BS CP3 *Code of basic data for the design of buildings* 'Sunlight: houses, flats and schools only'. London. British Standards Institution. 1945.
10 Petherbridge, P. *Sunpath diagrams and overlays for solar heat gain calculations* Building Research Establishment Current Paper, Research Series 39. 1965.
11 *Windows and environment* Pilkington Brothers Limited. 1969.
12 Cuttle, C. and Slater, A. I. 'A low energy approach to office lighting' *Light and Lighting* January/February 1975.
13 *Estimating daylight in buildings*—1, Building Research Station Digest 41. London. HMSO. 1970.
14 Hopkinson, R. G. *Architectural physics: lighting* London. HMSO 1963.
15 *Graphic design methods for natural lighting* Environmental Advisory Service Report. Pilkington Brothers Limited Not yet published.
16 The Illuminating Engineering Society *The IES Code for interior lighting* London. 1977.
17 British Standards BS CP3 *Code of basic data for the design of buildings* 'Daylighting'. London. British Standards Institution.
18 *Colour discrimination and heat rejecting window glasses* Internal report no 10/73. Plymouth Polytechnic School of Architecture.
19 Cuttle, C. *The use of special performance glazing materials in modern offices* MA Thesis. University of Manchester. 1974.
20 British Standards BS 3489 *Specification for sound level meters (industrial grade)* London. British Standards Institution. 1962.
21 British Standards BS 2475 *Specification for octave and one-third octave band-pass filters* London. British Standards Institution. 1964.
22 Kosten, C. W. and Van Os, G. J. 'Community reaction criteria for external noises' *The control of noise* NPL Symposium no 12 London. HMSO. 1962.
23 Parkin, P. H., Purkis, H. J., Stephenson, R. J. and Schlaffenberg, B. *London noise survey*. HMSO. London. 1968.
24 *Noise and buildings*. Building Research Station Digest (Second Series) no 38. 1963.
25 *Solar heat gain through windows* Environmental Advisory Service Report. Pilkington Brothers Limited. 1974. Fourth edition.
26 The Institution of Heating and Ventilating Engineers. *IHVE Guide Book A*. London. 1970.
27 *Solar control performance of blinds* Environmental Advisory Service Report. Pilkington Brothers Limited. July 1973.
28 Owens, P. G. T. and Barnett, M. *Building Services Engineer*, February 19 1974, pp250-252, and March 1974, p276.
29 Burberry, P. 'Conserving energy in buildings' AJ 11.9.74 p626.
30 *The building (second amendment) regulations* 1974 *statutory instruments no* 1944. HMSO.
31 Jennings, R. and Wilberforce, R. R. 'Thermal comfort and space utilisation' *Insulation* March 1973, pp57-60.

Part I Window Design

2 Worked example

Introduction

As previously explained in Chapter 1 windows cannot be designed in isolation from the rest of the building: many of the decisions in the design of the whole building are made, quite rightly, without much consideration of their effects on the windows but they do have effects that must be taken into account.

This chapter does not seek to establish any particular design system. It merely describes each step in the logical progress of the design of an office and shows how the facts that contribute to decision-making are acquired and how a decision at one step may modify previous decisions and affect later ones. The steps are not in the same order as the Design Guide: each building has its own characteristic priorities that determine the order in which the components of the design are considered. The detailed calculations that are used to illustrate many of the steps would, in practice, need to be repeated with each modification of the design. None of these repeats is calculated here and it is because of this lack of detailing in the interactions mentioned that this cannot be truly called a worked example.

Three intermingled streams of information flow are identifiable in the system described here and an attempt has been made to separate them in the layout. Column 1 contains *generalities*, discussions of the general influences on the design process. Column 2 contains *information* that is needed in the design. It may be of two sorts:

- Information that already exists and must be sought out. In this category are the recommendations and regulations that apply to the building, published research and the opinions of others, client's requirements, and the physical facts of the site.
- Information that must be generated either by calculation or by decision during the design process. These several aspects are exemplified by the consideration of daylight factor. There is likely to be a recommendation that a certain value of daylight factor is to be achieved in the building (existing information) so it is necessary to calculate that the expected daylight factor will be sufficient (calculated information) but in order to complete that calculation is it necessary to decide what room decorations will be used so that the reflectances can be fed into the calculation (decisive information).

Column 3 contains *decisions* which are also of two types:

- Contributive decisions, like that on room decorations, that are necessary before the design can be advanced to its next stage.
- Conclusive decisions that arise from the consideration of the preceding information input.

Reference is made to the relevant paragraphs of Chapter 1 and, where that does not give adequate information, to other items that influence the design. The prediction and calculation methods used in the examples are chosen with no intention of defining which of the many available methods should be used; to exemplify them all would only confuse.

GENERALITIES *INFORMATION/METHODS* *DECISIONS*

1 The client

1.01 Some decisions have already been made before the architect is engaged. The decision to have a building, or at least to consider having one, makes for the foundation of the brief from the client. This will describe the purpose of the building, the occupants and the available site. There may be an intermediate stage in which the architect, in association with appropriate consultants, undertakes very detailed studies that yield information for the formulation of the brief, such as details of movements of goods and/or people to establish the links between indoor spaces.

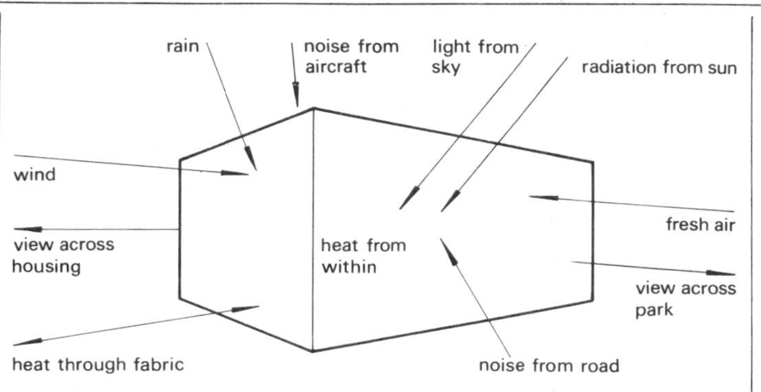

The architect's brief

1.02 An office, for the processing of bank records, to accommodate 200 people, office machinery and a computer. General office spaces may be 'open plan' but there must be 10 private offices and 3 executive offices. Office hours will be 0900-1700h on Monday to Friday. The site is rectangular (50 × 20 m) with a 50 m frontage onto a busy arterial road that runs from NE to SW and is 3·2 km from the perimeter of a major provincial airport, under the take-off flight-path. The latitude is 52°N. The building is to be designed to keep running costs to a minimum.

2 The Architect's role

2.01 The introduction to the Window Glass Design Guide considered that many architects begin the design process with some already-formed concept that arises from the brief, the site, and the character of the neighbourhood. This preliminary concept will include his first decisions on the appearance of the outside of the building: the ratio of solid to void in the facades, horizontal or vertical emphasis, deep construction or slab-type block, whether the building is to merge with its surroundings or stand out from them.

2.02 The space required for 200 people and the office machinery is 3000 m². (See *AJ metric handbook* section 12, p68.) Three floors will be needed to achieve this space on a site area of 1000 m². The computer can be housed in a basement and the remaining basement space can be used for car parking.

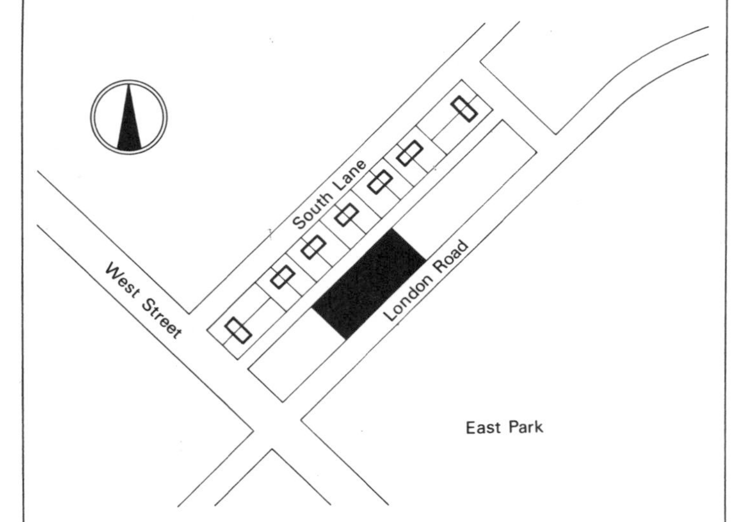

2.03 *Preliminary building concept: A rectangular block filling the site with a basement and 3 floors, a floor-to-floor height of 3·5 m and a glazed area of 50 per cent of the perimeter area. There will be vertical emphasis in the appearance of the front facade.*
For economy, the building will not be air conditioned and all glass will be above the working plane (0·85 m) so as to make the most efficient use of daylighting. The floor-to-ceiling height will be 2·5 m.

2.04 An attempt to sketch the building shows that, because the adjacent sites are built up to the boundaries the shorter facades cannot be glazed. The restrictions imposed by the shape of the site and the floor-to-ceiling height limit the glazed area to 34 per cent of the perimeter area if all the glass must be above the working plane. Vertical mullions of sufficient substance to give the required emphasis in appearance will reduce the glass area to 28 per cent or less. If the mullions are limited to 10 per cent of the glazing area and the floor-to-ceiling height can be increased by 500 mm, either at the expense of the ceiling void or by increasing the floor-to-floor height, the glass/perimeter ratio can be increased to 40 per cent.

2.05 *Decision: reduce glass/perimeter ratio from 50 per cent to 40 per cent; increase floor-to-floor height to 4·0 m; limit vertical mullions to 10 per cent of the glazing area.*

3 View out*

3.01 The view-providing function of windows is discussed in Chapter 1, paragraphs **1.01** to **1.08**. The following questions must be answered. Will the ability to see out of the building be important to the occupants? What are the characteristics of the views from the two glazed facades? What types of windows and types of glass, are needed to exploit the views to the best advantage? Does the same treatment have to be used on the two glazed facades?

3.02 The client will be consulted on the importance of view. His opinion may be based on former attitude surveys among the staff or it may be necessary to undertake such studies.

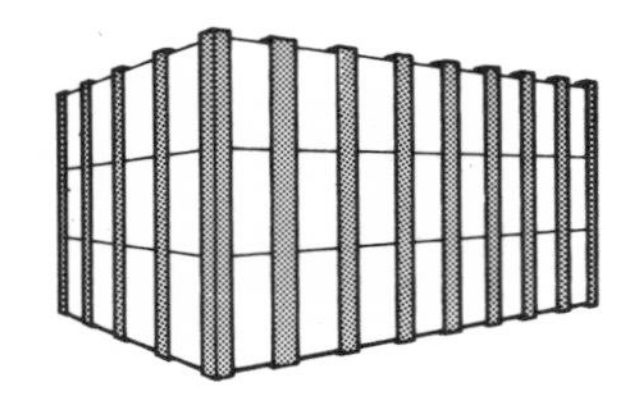

3.03 The client expects the staff to have the opportunity to see out of the building for relief from their close visual tasks. The view from the SE facade is across the road to open parkland which is unlikely to be developed. The view to the NW is across a narrow access road to a residential area of two-storey semi-detached houses. When the glazed area is not restricted to a small fraction of the facade (say, not less than 30 per cent) the shape of individual windows is of minor importance to the view-providing function. However, especially in open-plan offices, wide or deep mullions are very obstructive to an oblique view. Narrow mullions therefore contribute to the maintenance of the required glazed area and to the availability of a view.

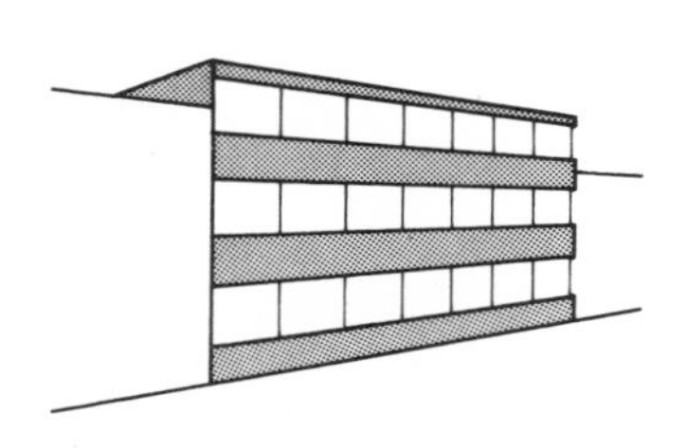

3.04 *Decision: keep mullion dimensions to a minimum: this largely destroys the opportunity to produce vertical emphasis in the facades and entails a change to horizontal emphasis. The NW and SE facades will have the same details.*

* See Chapter 1, Function 1, page 1.

4 Task illumination*

4.01 The requirement in the brief that running costs shall be kept to a minimum gives great importance to the decisions that are to be made about the means of lighting the interior of the building. The preliminary concept was a building with large windows so that a good deal of daylight would be available for task lighting. It is, however, necessary to calculate the extent to which daylighting can serve as task lighting and what supplementary artificial lighting will be needed. The first step is to calculate the daylight factors at various points on the working plane.

* See Chapter 1, Function 3, page 4.

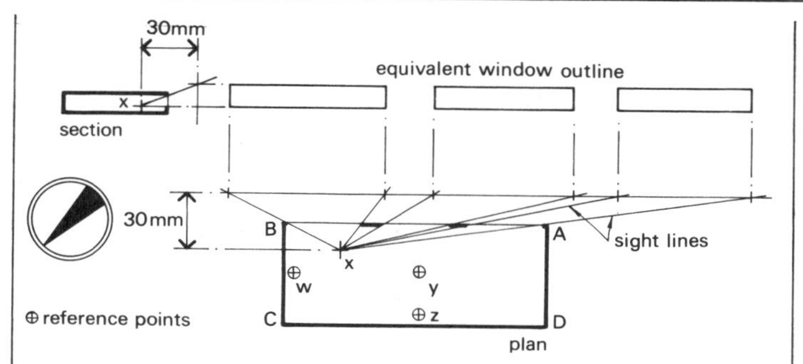

Daylight factor calculation

Sky component

4.02 1 Prepare a plan and vertical section of the area to be studied such that all the windows that may contribute to the daylighting are included.

2 Note the height of the working plane (0·85 m) and choose a series of reference points sufficient to determine the daylight distribution. Mark these on the plan and vertical section.

3 Draw sight-lines from a selected reference point to the limiting boundaries of all the glazing that can influence the daylight reaching that point.

4 Draw the normal from the reference point to each glazed facade and mark a distance of 30 mm from the reference point towards the glazing. Note that the 30 mm is a real distance and is not to be marked in the scale of the drawing.

5 At the 30 mm point, draw a line parallel to the glazing so that it is intercepted by the sight-lines drawn at stage 3.

6 Perform the operations of stages 3, 4 and 5 on both plan and vertical section and combine the two sets of intercepts to form the 'equivalent window outline' for use with the prediction methods based on gnomonic projection.

4.03 The geometric construction of the equivalent window outline has been described in detail but, in practice, it can frequently be simplified. For example, in a situation where mullions and glazing bars are slender, it is sufficient to take sight-lines to the boundaries of the whole glazed area, on plan and vertical section, and to calculate the ratio of clear glass area to total glazed area as a glazing bar factor to correct the subsequently calculated daylight factor.

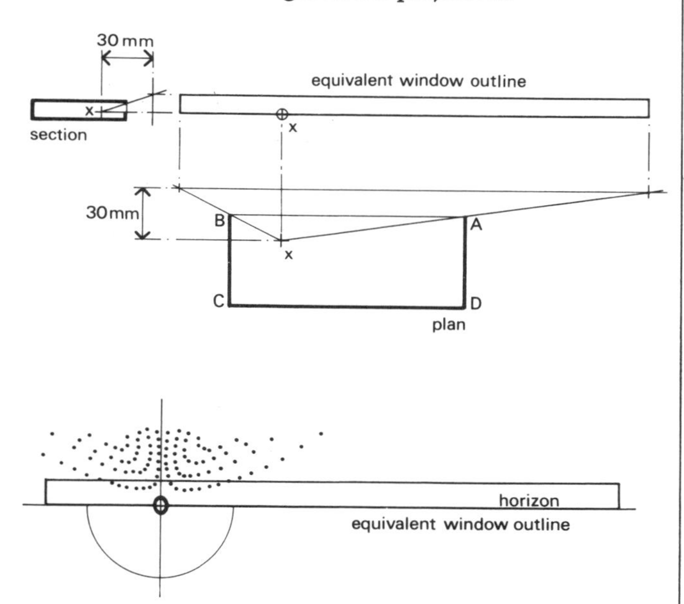

4.04 7 From part 6 of *Windows and environment* select overlay 3.1 (uniform sky conditions, overlay 3.2, are now rarely used for daylighting design) and superimpose it on the equivalent window outline so that its origin coincides with the position of the reference point and the 'horizon' line is horizontal. Count the dots enclosed by the window outline. Calculate the sky component (SC) by dividing the number of enclosed dots by 10.

4.05 *Decision: decide CIE sky (overcast).*

4.06 For reference point X:

Number of dots enclosed by window outline = 14

Sky component = 14 ÷ 10 = 1·4 per cent.

Repeat stages 3 to 7 for each of the other reference points.

4.07 The outlines of any obstruction to be seen through the windows from the reference point may be drawn by the same geometric construction as used for the equivalent window outline. Any dots enclosed by the outlines of obstructions would not be counted as part of the sky component but would constitute an *externally reflected component* (ERC) calculated by dividing the number of enclosed dots by 10, as for the sky component, and multiplying the result by a factor that takes account of the reflective properties of the facade of the obstruction.

Internally reflected component

4.08 Calculate the parameters of the internally reflected component (IRC) by the method described in *Windows and environment*, part 2, section 4.

1 Calculate: room index $= \dfrac{W \times L}{h\,(W + L)}$, where

W = width of room parallel to window wall;
L = length of room, perpendicular to window wall;
h = height from working plane to ceiling.

4.10 W = 50 m; L = 20 m; h = 2·15 m [3·0—0·85].

$$\text{Room Index} = \frac{50 \times 20}{2{\cdot}15\,(50 + 20)} = \frac{1000}{2{\cdot}15 \times 70} = \frac{1000}{150{\cdot}5} = 6{\cdot}6$$

2 Calculate: glazing/floor area ratio = g/f, where
g = area of the glazing; f = area of the floor.
Area of glass = 50 × 2 × 2 = 200 m²
Area of floor = 50 × 20 = 1000 m²
Glazing/floor area ratio = 200/1000 = 0·20.

3 Find the values of three factors (a, v and e) from tables 4.1, 4.2 and 4.3 of *Windows and environment*. (See Tables I, II and III below.) Factors a and v depend upon the room index and on the reflectances of the floor, ceiling and walls.

Table I Values of a

	reflectance										
Floor	0·3			0·1							
Ceiling	0·7			0·7			0·5			0·3	
Walls	0·5	0·3	0·1	0·5	0·3	0·1	0·5	0·3	0·1	0·3	0·1
Room index	*Values of* **a**										
1·0	1·1	1·0	1·0	1·1	1·0	1·0	1·0	1·0	1·0	1·0	1·0
1·25	1·2	1·1	1·0	1·1	1·1	1·0	1·0	1·0	1·0	1·0	1·0
1·5	1·3	1·2	1·1	1·3	1·2	1·0	1·1	1·0	1·0	1·0	1·0
2·0	1·4	1·3	1·2	1·4	1·3	1·1	1·1	1·1	1·1	1·1	1·0
2·5	1·6	1·5	1·3	1·5	1·4	1·2	1·2	1·2	1·1	1·1	1·0
3·0	1·8	1·6	1·4	1·6	1·5	1·2	1·3	1·2	1·1	1·2	1·1
4·0	2·1	1·9	1·6	1·9	1·7	1·4	1·4	1·3	1·2	1·3	1·2
5·0	2·4	**2·2**	1·7	2·1	1·9	1·5	1·6	1·5	1·3	1·5	1·2

Table II Values of v

	reflectance										
Floor	0·3			0·1							
Ceiling	0·7			0·7			0·5			0·3	
Walls	0·5	0·3	0·1	0·5	0·3	0·1	0·5	0·3	0·1	0·3	0·1
Room index	*Values of* **v** *(CIE sky)*										
1·0	3·9	3·2	1·4	3·2	2·4	1·3	2·2	1·8	1·0	1·4	0·8
1·25	3·8	3·1	1·4	3·1	2·3	1·3	2·1	1·7	1·0	1·4	0·7
1·5	3·7	3·0	1·3	3·0	2·2	1·2	2·0	1·7	1·0	1·3	0·6
2·0	3·5	2·8	1·3	2·8	2·0	1·1	1·9	1·6	0·9	1·2	0·4
2·5	3·3	2·6	1·2	2·6	1·8	1·0	1·8	1·4	0·8	1·1	0·4
3·0	3·1	2·4	1·1	2·4	1·7	1·0	1·7	1·3	0·8	1·0	0·3
4·0	2·6	2·0	1·0	2·0	1·3	0·8	1·5	1·1	0·7	0·8	0·3
5·0 +	2·2	**1·5**	0·9	1·6	1·0	0·7	1·2	0·9	0·6	0·6	0·3
Room index	*Values of* **v** *(uniform sky)*										
Any	4·2	3·5	1·8	3·4	2·6	1·3	2·6	1·8	1·0	1·5	0·9

Table III Value of e

∝	e	h/d	e
0	**1·00**	0	1·00
10	0·88	0·2	0·87
20	0·74	0·4	0·71
30	0·56	0·6	0·55
40	0·39	0·8	0·42
45	0·31	1·0	0·31
50	0·23	1·5	0·14
60	0·10	2·0	0·07

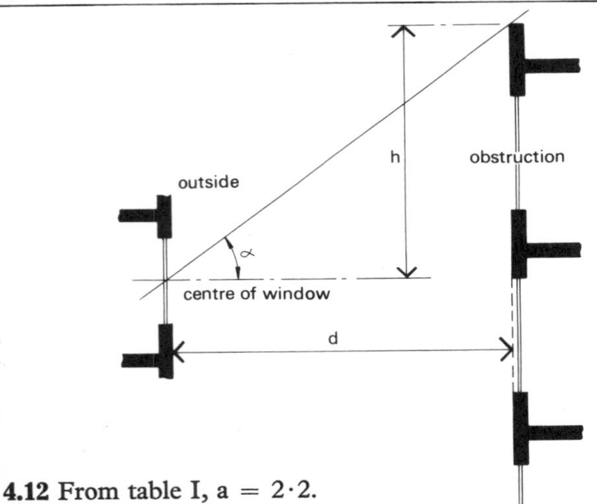

4.12 From table I, a = 2·2.
From table II, v = 1·5.
From table III, because there is no obstruction, e = 1·0.

4.09 *Decision: The height of the ceiling void from ceiling to the floor above will be 1 m to accommodate any ventilation ducts that may be required. The office will be open-plan and occupy the whole of the floor area.*
The useful height of the window will be 2m.

4.11 *Decision: Room decoration. The ceiling will have light finish, reflectance = 0·7.**
The walls will have a coloured finish of medium lightness, reflectance = 0·3.
The floor will be carpeted in a light colour, reflectance = 0·3.

* For reflectances, see *Windows and environment*.

Daylight factor

4.13 Calculate the uncorrected daylight factor (DF) by combining the elements of the calculation in the formula:

DF = a (SC + ERC) + veg/f

DF = 2·2 (1·4 + 0) + 1·5 × 1·0 × 0·2

DF = 3·38 per cent.

The daylight factor so calculated is that at the reference point X due to the windows in the facade AB. For an open-plan office the sky component in the calculation would be increased by the sky component at X due to the windows in the facade CD. The resulting total daylight factor is for windows glazed with absolutely clean, 6 mm clear glass. In practice the value will probably need to be modified by factors to take account of differences in the visible transmittance of the glass and the effect of dirt. The tabulated values of maintenance factor assume that the glass has collected dirt to the point at which it would normally be cleaned. (See Table IV.)

Table IV Maintenance factor

Location	*Maintenance factor, M*		
	Glazing		
	Vertical	*Sloping*	*Horizontal*
Clean	0·9	0·8	0·7
Industrial	0·8	0·7	0·6
Very dirty	0·7	0·6	0·5

4.14 Sky component from facade CD = 0·2 per cent

Glazing bar factor, B = 0·98

Maintenance factor, M = 0·9.

DF = BM [a (SC + ERC) + veg/f]

DF = †0·98×0·9 [2·2 (1·4+0·2+0) + 1·5×1·0×0·2]

DF = 0·98×0·9 (3·52+0·3) = 0·98×0·9×3·82 = 3·37

DF = 3·4 per cent.

Calculate the daylight factor for each of the remaining reference points, using the values of sky component discovered at stage 7 (see 4.04).

†Two per cent of window area ie 0.98.

4.15 The details of the daylight factor distribution are used to assess the extent to which daylighting will be adequate. At the central reference point, Y, the daylight factor is 1·7 per cent. This is a little less than the 2 per cent recommended as a minimum for general offices but, since it is likely that the central band of the office floor will be used for communications, it is reasonable to conclude that, with clear glazing, daylighting will be adequate for most of the working day and artificial lighting will be required for early mornings and late afternoons in winter.

4.16 *Decision: accept that daylighting will be generally adequate. Provide artificial lighting to a level of* 500 *lux on the working plane (from* 4.15).*

* See *The IES code for interior lighting*. The Illuminating Engineering Society. London. 1977. Page 57. Also British Standard Code of Practice CP 3: Chapter 1: Part 1 *Code of basic data for the design of buildings*. London. 1964. Page 27.

5 Sunlight penetration

5.01 This stage studies the admission of sunshine to the rooms to see that there is sufficient to give a pleasant appearance and a feeling of well-being and not enough to cause visual discomfort by glare: the effect of sunshine on the thermal balance of the building is studied separately at a later stage. The method described is to take each of the reference points selected for the study of daylight quantity and use the graphic methods of *Windows and environment* to show the periods when the sun can shine onto them. The example is for the penetration of sunlight through the south-east windows to the reference point X. See Chapter 1, Function 2, page 2.

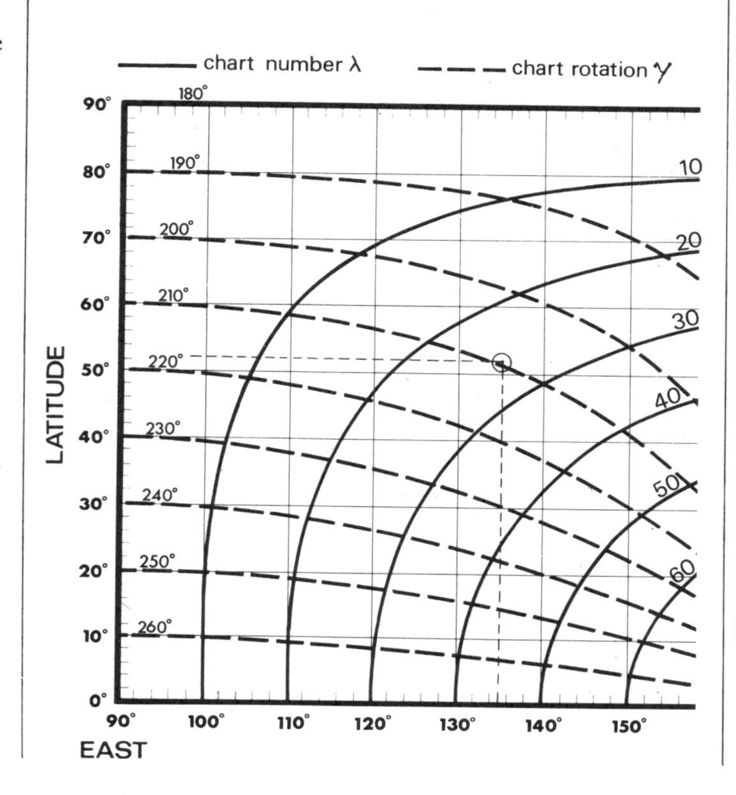

5.02 1 From the drawings of equivalent window outline prepared for the study of task illumination (stage 4) select that for reference point X and the south-east facade and draw on it a vertical line through X.

2 From *Windows and environment*, part 6, select the sun chart for latitude 52°N (the latitude of the site) and for azimuth 135° (the direction of the normal to the outside of the window wall) and find the chart rotation angle, γ. The chart number and chart rotation are found from the sun chart selector diagram inside the front cover of the folder of overlays. Part of the diagram, with this example added, is shown in the figure on previous page. Locate the required point by reference to the rectangular grid, marked in latitude angles up the left side and in azimuth angles along the bottom, then read the required numbers by interpolating in the grid of curved lines:

Chart number, $\lambda = 26$

Chart rotation, $\gamma = 210°$

Calculate $(360\text{-}\gamma) = 150°$.

3 Select sun chart number 26 and superimpose it upon the equivalent window outline so that the central '0' of the chart coincides with the position of the reference point, X. Rotate the chart until the 150° mark (360-γ) on the perimeter circle of the chart is vertically below the reference point.

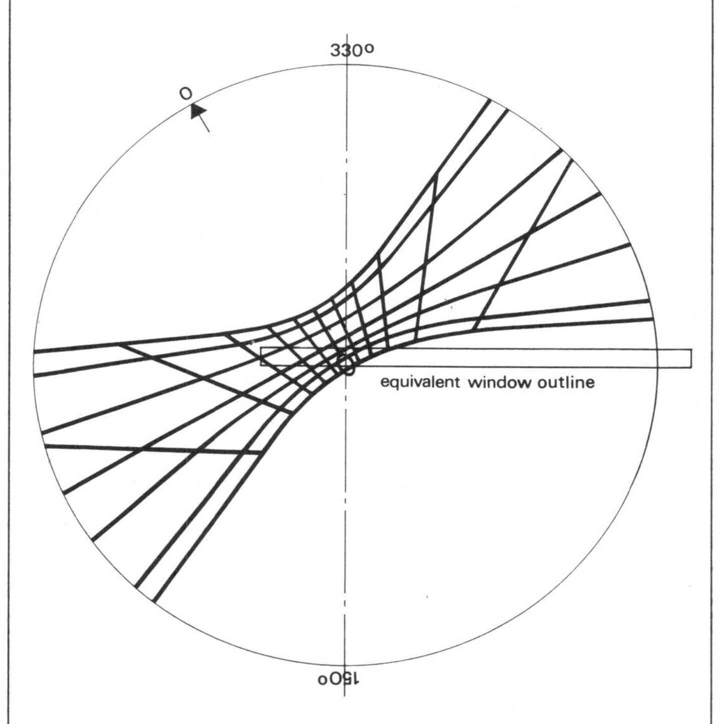

4 Label the season lines and time lines. In the northern hemisphere the winter solstice occurs in December. This is represented by the lowest sun path in the diagram and the other sun paths are labelled in sequence from it. The equinox sun path crosses the horizon at 0600h and the time lines can be labelled in sequence from that point.

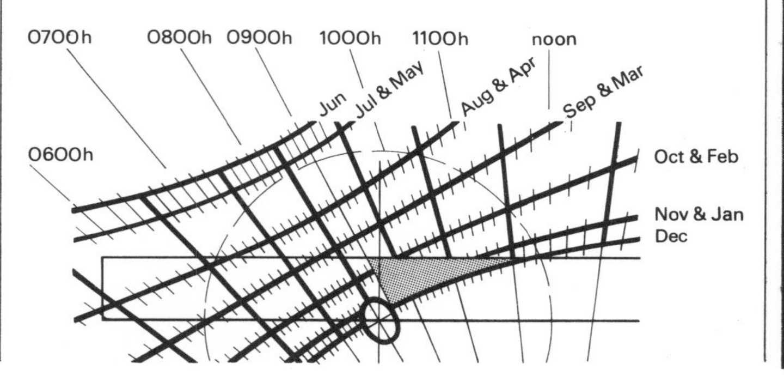

5 Examine the periods of sun penetration to the reference point. The significant region is shaded in the figure. It lies outside the period of daylight saving so that the solar times marked coincide with the practical time scale, GMT. The office is not occupied before 0900h, so the longest period during which it is possible for the sun to shine onto a worker at the point, X, is 2¾ hours from 0900-1145 in mid-winter. The probable total of bright sunshine during this period is 0·625 hours (see BRE Current Paper 75/75. *Availability of sunshine* Figure 4). It would hardly be necessary to provide shading devices to protect the point X from sun glare for a probable daily period of ½ hour: points nearer to the SE facade than X will experience longer periods of insolation and the decision on shading will depend upon the conditions at points close to the windows. The inner circle on the sun chart has a radius of 30 mm and represents directions at an angle of 45° to the direction of view. Experience has shown that glare sources outside this circle are of minor importance so, for a direction of view normal to the windows, the risk of glare problems is restricted to the first 2½ hours of the working day and the greatest probable total of bright sunshine during this period is found to be 1 hour 9 minutes, occurring in October.

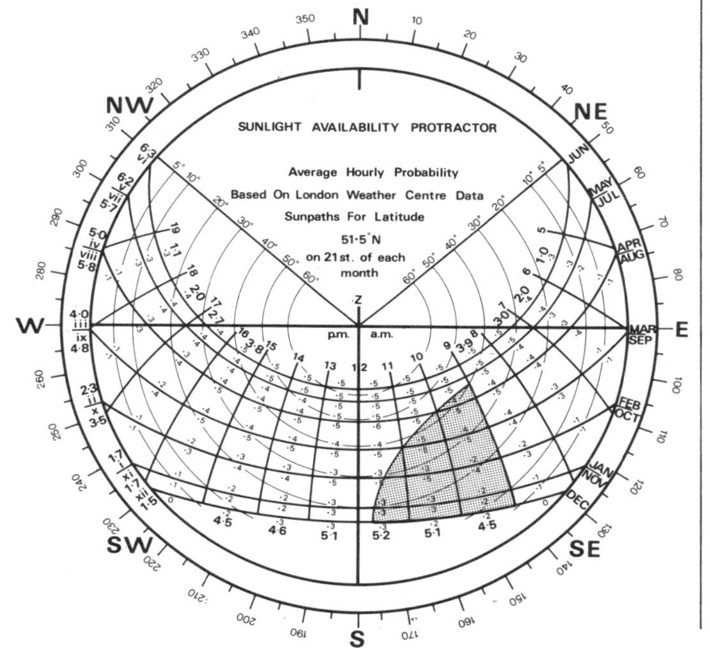

5.03 *Decision: Sunlight penetration is adequate to give the office a pleasant appearance. Position desks so that the general direction of view is easterly and there will be no need for shading to combat sun glare.*

6 Sound insulation

6.01 Having checked the proposed window configuration and glazing for view out, task illumination, and sunlight penetration, the next step is to check for sound insulation against external noise (street traffic and aircraft). The method has been described in Chapter 1, Function 6, page 12 and is summarised here.

Method

6.02 Three decisions/computations must be made:

- Calculate or measure external noise climate
- Establish the acceptable internal noise climate
- Decide on a window and glazing type with the required acoustic insulation value to modify external noise levels.

6.03 Chapter 1, Function 6, page 12, described two methods for this three-stage computation: a complex and relatively precise method (paragraph **6.14, 6.15**) and a simpler less precise one (paragraph **6.16, 6.17**).

6.04 Final selection (particularly in critical situations of high external noise levels and stringent internal requirements) is best made using the more complex method; but it will not be employed here because a great deal of esoteric, inaccessible information is required (eg sound spectra for specific aircraft). It is assumed that in such cases specialist assistance will be called in.

6.05 Thus the following worked example uses the simpler method, which is useful for general guidance and initial decision-making, and easily operable by ordinary non-specialist practitioners.

Example

Road traffic noise: south-east facade

6.06 From Table V, establish *external* noise climate—that level of sound intensity which is exceeded for 10 per cent of the time between the hours of 0800 and 1800, measured in dBA.

Table V Noise levels exceeded for 10 per cent of the time at various locations

Group location		10 per cent noise levels—dBA 0800-1800 hours	0100-0600 hours
A	Arterial roads with many heavy vehicles and buses (kerbside)	80	68
B i	Major roads with heavy traffic and buses	75	61
ii	Side roads within 15-20 m of A or Bi group roads		
C i	Main residential roads		
ii	Side roads within 20-50 m of heavy traffic routes	70	54
iii	Courtyards of blocks of flats screened from direct view of heavy traffic		
D	Residential roads with local traffic only	65	52
E i	Minor roads		
ii	Gardens of houses with traffic routes more than 100 m distant	60	48
F	Parks, courtyards, gardens in residential areas well away from traffic routes	55	46
G	Places of few local noises and only very distant traffic noise	50	43

- The building facade is at the kerbside of an arterial road with many vehicles and buses (group A on table), therefore relevant noise level is 80 dBA.

6.07 From Table VI, establish acceptable *internal* noise level, again measured in dBA (to allow valid comparison with previous figure).

Table VI Acceptable noise levels for various situations

Situation	dBA
Large rooms for speech such as lecture theatres, large conference rooms or council chambers	30
Bedroom in urban area	35
Living room in country area	40
in suburban area	45
in busy urban area	50
School classroom	45
Executive office	45-50
General office	55-60

● For general offices, 55 to 60 dBA would be acceptable; for executive offices, 45 to 50 dBA.

6.08 From Table VII, establish the required *sound insulation* of the windows on the south-east facade to obtain the necessary reduction of road traffic noise.

Table VII The required sound insulation of windows for various reduction of road traffic noise

Difference between outdoor and indoor levels—dBA	Window insulation (100-3150 Hz)dB
20-25	20-25
28	30
30	35
35	40
40	45

● For general offices, the difference between outdoor and indoor levels is (80 minus 55) to (80 minus 60),* therefore 20 to 25 dBA. From Table VII, required insulation is therefore 20 to 25 dB.

● For executive offices, the difference between outdoor and indoor levels is (80 minus 45) to (80 minus 50), therefore 30 to 35 dBA. From Table VII, required insulation is therefore 35 to 40 dB.

Road traffic noise: north-west facade

6.09 From table V, establish *external* noise climate as before. Traffic in South Lane is very light (and greatly shielded by intervening houses), and there are only occasional vehicles in the access road.

● The appropriate category is D, and noise level is 65 dBA. As road traffic noise is relatively low at this facade, air traffic noise might well be the major noise problem to influence choice of glazing type and window design; therefore aircraft noise will be investigated before carrying out computation to determine required insulation value of window.

Air traffic noise: north-west facade.

6.10 The building is under the take-off flight path of an airport 3·2 km away;† direct sound measurements are the most reliable method of establishing air traffic noise, using a sound level meter measuring in dBA. Road traffic noise must be excluded from this measurement as far as possible; so the readings are taken at a position near the site, well screened from the main arterial road by existing buildings, and more than a metre from any hard sound-reflecting surface. Measure first the *general* noise level when no aircraft can be heard; then the *peak* combined noise level when an aircraft passes overhead.

● Peak air traffic noise level is 75 dBA.

6.11 Clearly, air traffic noise (at 75 dBA), not road noise (at 65 dBA), is the major problem to be catered for at the north-west facade (whereas at the south-east facade it is road traffic noise that creates the major problem, at 81 dBA).

6.12 From table VI, establish acceptable *internal* noise levels, as before:

● For general offices, 55 to 60 dBA; for executive offices, 45 to 50 dBA.

6.13 From table VII, establish the required *sound insulation* of the windows on the north-west facade to obtain the necessary reduction of air traffic noise.

* 80 being the outdoor figure, and 55 to 60 the range of indoor figures.

† See original brief, paragraph 1.02, page 31.

• For general offices, the difference between outdoor and indoor levels is (75 minus 55) to (75 minus 60),* therefore 15 to 20 dBA. From Table VII, required insulation is therefore 20 dB.
• For executive offices, the difference between outdoor and indoor levels is (75 minus 45) to (75 minus 50), therefore 25 to 30 dBA. From Table VII, required insulation is therefore 30 dB.

Choice of window and glazing types
6.14 Table VIII gives approximate typical insulation values for various window and glazing types.

Table VIII Typical insulation values for windows

Description	Approx sound insulation†
Wide open window	5-10 dB
Slightly open window	10-15 dB
Closed window (capable of being opened)	15-20 dB
Sealed single-glazed window, 4 mm	20-25 dB
Staggered opening double-glazed window	ditto
Sealed single-glazed window, 6 mm	25-30 dB
Well fitting double-glazed window, 4 mm	ditto
Sealed single-glazed window, 10 mm	30 dB
Well fitting double-glazed window, 6 mm	30-35 dB
Sealed double-glazed window, 4 mm	35-40 dB
Sealed single-glazed window, 25 mm	ditto
Sealed double-glazed window, 6 mm	40-45 dB
Sealed double-glazed window, 4 mm, one leaf, 6 mm the other	ditto
Sealed double-glazed window with leaves out of parallel, 10 mm one leaf, 12 mm the other	45-50 dB

† Average value over 100 to 3150 Hz frequency range

• To obtain 20 dB sound insulation (general offices), north-west facade), an ordinary single-glazed window will do—but only so long as it is closed; if it is opened for ventilation, noise insulation performance drops disastrously.
• To obtain 20 to 25 dB sound insulation (general offices, south-east facade) a sealed single-glazed window will be required (ie no natural ventilation), or double-glazed opening windows which are staggered.
• To obtain 30 dB sound insulation (executive offices, north-west facade), a sealed 10 mm single-glazed window (ie no natural ventilation) will be required.
• To obtain 35 to 40 dB sound reduction (executive offices, south-east facade), a sealed double-glazed window or sealed single-glazed window of 25 mm plate is required.
6.15 Clearly there can be no opening windows for natural ventilation on either facade if the windows are to adequately perform their function as sound filters against external noise. Therefore facades must be sealed; and to allow for a margin of safety it is decided to provide all executive offices with double glazing, while single glazing will perform satisfactorily for general offices.
6.16 So decision is as shown at right.

* 75 being the outdoor figure, and 55 to 60 the range of indoor figures.

6.17 *Decision. The facades will be sealed and mechanical ventilation will be provided. General offices will be single glazed. Private offices will be double glazed.*

7 Thermal control

7.01 Having considered the performance of the windows in terms of view out (paras **3.01** to **3.04** of worked example) daylight factor (paras **4.01** to **4.14**) sunlight penetration (paras **5.01** to **5.03**) and noise insulation (paras **6.01** to **6.16**) we now look at the proposed windows as heat filters.
7.02 This involves assessing their performance both in terms of *heat gain* and *heat loss*—see Chapter 1, Function 7, page 15.

See next page for paragraphs **7.03** *and* **7.04.**

Methods
7.05 Two initial computations need to be made on *external conditions:* peak outdoor temperature (both the time at which it occurs and its magnitude) and peak instantaneous heat gain through the windows (again, both the time when this occurs and the magnitude of the gain).
7.06 Then the *internal conditions* in the building can be deduced, on the basis of the external conditions established above, interacting with the proposed building design.
The method comprises the following steps:
Start with instantaneous outdoor temperature (from **7.07**), T_0. Add temperature rise inside building due to solar heat gain through windows, to give indoor temperature T_1. Add further temperature rise inside building due to heat from people and machinery, to give indoor temperature T_2. Add further temperature rise due to heat from lights, to give final indoor temperature T_3.
If internal conditions are found to be unsatisfactory, design must be amended (with the aid of rapid computer calculations to evaluate alternative options) until satisfactory conditions are ensured.

7.03 The thermal design of the windows is based on the methods recommended by the Institution of Heating and Ventilating Engineers and described in the *IHVE guide*, book A.[26]

7.04 The first step is to identify the seasons and times that are expected to give the most difficulty in terms of *heat gain*, and use short-cut methods of calculation to reach a design solution. More sophisticated methods can then be used, if necessary, to study the component parts of the design in detail and make any adjustments.

Now turn back to paragraph **7.05.**

Example

External conditions

7.07 Use of one of the available computer programmes (or, if a manual calculation is done, Table A8.3 from the *IHVE guide*) will indicate that the highest outdoor temperature occurs at 1500h in July, and its magnitude is 25·5°C.

Table A8.3. Typical average outdoor air temperatures for 5% of days of highest solar radiation. (51·7°N latitude).

Month	Daily mean t'_{ao} (°C)	Typical temperatures, hour by hour (°C) — Sun time 0600	0700	0800	0900	1000	1100	1200	1300	1400	1500	1600	1700	1800	Mean to peak swing $\tilde{t}_{ao}$ (°C)
March	7·0	2·8	4·0	5·4	7·0	8·6	10·0	11·2	12·2	12·8	13·0	12·8	12·2	11·2	6·0
April	9·0	4·8	6·0	7·4	9·0	10·6	12·0	13·2	14·2	14·8	15·0	14·8	14·2	13·2	6·0
May	13·0	8·1	9·5	11·2	13·0	14·8	16·5	17·9	19·1	19·8	20·0	19·8	19·1	17·9	7·0
June	16·5	11·2	12·8	14·6	16·5	18·4	20·2	21·8	23·0	23·8	24·0	23·8	23·0	21·8	7·5
July	19·0	14·4	15·7	17·3	19·0	20·7	22·3	23·6	24·6	25·3	25·5	25·3	24·6	23·6	6·5
August	17·0	12·8	14·0	15·4	17·0	18·6	20·0	21·2	22·2	22·8	23·0	22·8	22·2	21·2	6·0
September	13·5	9·6	10·7	12·1	13·5	14·9	16·3	17·4	18·3	18·8	19·0	18·8	18·3	17·4	5·5

7.08 Use of the computer will then also indicate the times when the amount of solar energy falling upon the two main facades of the building (south-east and north-west) is at maximum, and will indicate the intensities involved. If, alternatively, a manual calculation is done, the figures can be most easily be found from *Solar heat gain through windows*.*[25]

7.09 Select graph 2 for 52°N latitude, superimpose it on graph 1, and adjust until the arrow indicates a window orientation of 135° for the south-east facade.

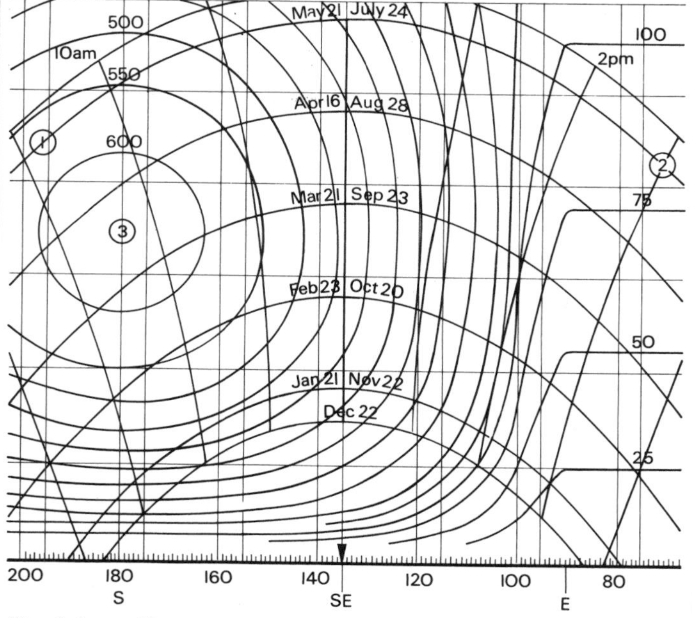

Read from diagram:

1 the peak instantaneous gain for the month of July (570 W/m², by interpolation between 550 and 600 W/m², on the May 21/July 24 curve)

2 the gain for 1500h in July, when the outside air temperature is highest (80 W/m², by interpolation between 75 and 100, where the curve for 1500h intersects that for May 21/July 24)

3 the highest value of heat gain during the year (630 W/m²).

7.10 Reset arrow to 315° for the north-west facade, and read off corresponding data.

Table IX Heat gain through windows

Facade	Period	Month	Time	Heat gain
South-east	Peak heat gain	April	0930	630 W/m²
,,	High outdoor temperature	July	0915	570 W/m²
,,	Peak outdoor temperature	July	1500	80 W/m²
,,	Peak heat gain	August	0930	630 W/m²
North-west	Peak heat gain	June	1700*	450 W/m²
,,	Peak outdoor temperature	July	1500	180 W/m²
,,	High outdoor temperature	July	1700*	420 W/m²

* The peak value occurs at 1730h which is after office hours.

Internal conditions

7.11 The computer can now calculate the internal thermal conditions in the building from three sets of data:

* *Environmental Advisory Service report* Pilkington Brothers Limited. 1974. Fourth edition.

1 the external conditions established in the previous section
2 the internal sources of heat gain within the building (people, machinery and lights)
3 the nature of the building fabric (glass-to-solid wall ratio; depth of rooms behind windows; massiveness of the building, properties of the window glazing) and the number of air changes per hour.

7.12 The following decisions are fed in:

7.13 Heat gain from *people*: see *IHVE guide*, Table A7.1

Table A7.1. Heat emission from the human body. (Adult male, body surface area 2 m²)

Application			Sensible (s) and latent (l) heat emissions, W, at the stated dry-bulb temperatures, °C.									
			15		20		22		24		26	
Degree of activity	Typical	Total	(s)	(l)	(s)	(l)	(s)	(l)	(s)	(l)	(s)	(l)
Seated at rest	Theatre, hotel lounge	115	100	15	90	25	80	35	75	40	65	50
Light work	Office, restaurant*	140	110	30	100	40	90	50	80	60	70	70
Walking slowly	Store, bank	160	120	40	110	50	100	60	85	75	75	85
Light bench work	Factory	235	150	85	130	105	115	120	100	135	80	155

• If heat emission is 100 W per person at a dry bulb temperature of 20°C (see table), and if the building is occupied by one person for every 10 m² of overall floor area, the heat gain from people equals 10 W/m² of floor area.

7.14 *Heat gain from people equals* 10 *W/m² of floor area.*

7.15 Heat gain from *machinery*. This can be estimated from the detailed brief:

• 10 W/m² of floor area.

7.16 *Heat gain from machinery equals* 10 *W/m² of floor area.*

7.17 Heat gain from *artificial lights*. This can be derived from the *IHVE guide*, Table A7.6.

Table A7.6. Connected loads for lighting equipment.

Level of illumination (lux)	Approximate total requirement,* W/m² floor area							
	Tungsten		Mercury	Fluorescent				
				65 W white			65 W de luxe warm white	
	Open enamel industrial reflector (300 W)	General diffusing (200 W)	MBF industrial reflector (250 W)	Enamel or plastic trough	Enclosed diffusing fitting	Louvred ceiling panels	Enclosed diffusing fitting	Louvred ceiling panels
200	25–32	32–45	10–15	7·5	10	10	15–20	17–22
400	50–65	65–90	20–30	15	15–22	20	30–40	32–45
1000			55–75	32–45	42–55	45–65		

• If level of illumination on the working plane is 500 lux,* and fittings are 65 W white fluorescent tubes behind louvred ceiling panels then, from table, total wattage (by interpolation) = 25 W/m².

7.18 *Heat gain from electric lighting equals* 25 *W/m² of floor area.*

7.19 Nature of proposed building fabric, in terms of a shading/storage factor S, which takes account of two characteristics of the proposed building design—first, the transmissive and absorptive properties of the *glass;* second the thermal storage properties of the *building mass.*

• For glass properties, see Chapter 1, Function 8, Table XII, page 23. In the case of clear single glazing, $SWSC = 0 \cdot 92$ and $LWSC = 0 \cdot 05$; in the case of clear double glazing, $SWSC = 0 \cdot 74$ and $LWSC = 0 \cdot 10$.

• For thermal storage properties of building mass, see Chapter 1, Function 7, Table I, page 16. If structure is reinforced concrete of average density 2250 kg/m³, and all floors are covered with carpet tiles, storage factor, $FS = 0 \cdot 70$.

• For combined shading/storage factor,
$S = FS \times SWSC + LWSC$. In the case of single-glazed office space

$$S = 0 \cdot 70 \times 0 \cdot 92 + 0 \cdot 05 = 0 \cdot 644 + 0 \cdot 05 = 0 \cdot 694$$

In the case of double-glazed office space

$$S = 0 \cdot 70 \times 0 \cdot 74 + 0 \cdot 10 = 0 \cdot 518 + 0 \cdot 10 = 0 \cdot 618$$

7.20 *Factor* S *equals* 0·694 *for general offices and* 0·618 *for private offices.*

7.21 Nature of proposed building fabric, in terms of ratio of glass-to-wall area for the room, measured on the inside surface of the window wall.

* From brief.

• From an earlier decision in this worked example (para **4.09**, page 34) window height is 2 m clear, and wall height from floor to ceiling is 3 m (4 m floor-to-floor height minus 1 m ceiling space). Therefore ratio of glass-to-wall on inside surface is 2:3 = 2/3.

7.22 *Glass-to-wall ratio equals* 2/3.

7.23 Depth of room inwards from window. Because total depth of building is 20 m (plan shape is 50 m long × 20 m wide), and there will be some private offices within the general office space for executives, a decision must now be taken on the dimensions and arrangement of the private offices.

• If private offices are made 5 m deep, there will be three offices depths to be considered: 20 m, 15 m and 5 m.

7.24 *Room depths are* 20 *m*, 15 *m*, *and* 5 *m*.

7.25 Ventilation rate. This is taken at 10 air changes per hour (from *IHVE guide*, table A9.1).

7.26 *Ventilation rate is* 10 *air changes an hour.*

Calculation of internal conditions.

7.27 Now internal temperatures can be calculated, using the three sets of data identified in **7.11** and set out in subsequent paragraphs. The starting point is external air temperature for each of the chosen times, T_0, from the *IHVE guide* table A8.3. (See paragraph **7.07**).

7.28 Three sets of indoor temperature can now be calculated:
T_1 (the temperature that would result from adding, to T_0, the temperature rise due to solar heat gain, dT);
T_2 (the temperature that would result from the further addition of the rise due to people and machinery);
T_3 (the final temperature, resulting from the further addition of the rise due to electric lighting).

7.29 To calculate dT, due to solar heat gain, the following formula is used:

$$dT = \frac{IfS}{0{\cdot}33\,ND}$$ where dT is temperature rise, and:

I = instantaneous heat gain through unit area of glass (in W/m^2, from **7.10**)
f = glass-to-wall ratio of room (from **7.22**)
S = combined shading/storage factor (from **7.20**)
N = number of air changes per hour (from **7.26**)
D = depth of room inwards from window (from **7.24**)

7.30 Starting with the general office, 15 m deep, single glazed, the calculated internal temperature at the significant times of year are as shown in column T_3. As an example of how this is derived, the calculation for the top line of figures is this:

1 The highest value of heat gain during the year was found in **7.10** (630 W/m^2 at 09.30h in April). At this time

$$dT = \frac{IfS}{0{\cdot}33\,ND} = \frac{630 \times 0{\cdot}667 \times 0{\cdot}694}{0{\cdot}33 \times 10 \times 15} = \frac{291{\cdot}06}{49{\cdot}5} = 5{\cdot}87$$

2 Therefore $dT = 5{\cdot}9$ (5·87 rounded up).
3 T_0 is found as explained in **7.27** and **7.07**, and equals (in this case) 9·8°C (by interpolation on table A 8.3. Paragraph **7.07**).
4 Therefore $T_1 = 5{\cdot}9 + 9{\cdot}8 = 15{\cdot}7$°C (say 16°C).
5 Internal heat gain from people and machinery is 20 W/m^2 and this results in a temperature rise of 2°C (calculated by computer; detailed steps not shown here).
6 Therefore T_2 = 16°C + 2°C = 18°C (rounded figures).
7 Internal heat gain from lights is 25 W/m^2 and this results in a temperature rise of 2·5°C (calculated by computer; detailed steps not shown here).
8 Therefore T_3 = 18°C + 2·5°C = 20°C (the apparent arithmetic error is explained by the fact that both 18° and 20° are rounded-up figures).
Tables X to XII show calculated indoor temperatures T_3 for the three types of office which may occur.

* See tables on next page.

Table X General office, 15 m deep, single glazed

Facade	Month	Time	Gain W/m²	dT °C	To °C	T_1 °C	T_2 °C	T_3 °C
S-E	April	0930	630	5·9	9·8	16	18	20
	July	0915	570	5·3	19·4	25	27	30
	July	1500	80	0·7	25·5	26	28	31*
	August	0930	630	5·9	17·8	24	26	28
N-W	June	1700	450	4·2	23·0	27	29	32
	July	1500	180	1·7	25·5	27	29	32
	July	1700	420	3·9	24·6	28	30	33*

Table XI Private office, 5 m deep, double glazed

Facade	Month	Time	Gain W/m²	dT °C	T_0 °C	T_1 °C	T_2 °C	T_3 °C
S-E	April	0930	630	15·7	9·8	26	28	30
	July	0915	570	14·2	19·4	34	36	38*
	July	1500	80	2·0	25·5	28	30	32
	August	0930	630	15·7	17·8	34	36	38
N-W	June	1700	45¾	11·2	23·0	34	36	39
	July	1500	180	4·5	25·5	30	32	34
	July	1700	420	10·5	24·6	35	37	40*

Table XII General office, 20 m deep, single glazed, heat gain through both facades

Month	Time	Gain W/m²	dT °C	T_0 °C	T_1 °C	T_2 °C	T_3 °C
April	0930	710	4·9	9·8	15	17	19
June	1700	508	3·5	23·0	26	25	31
July	0915	658	4·6	19·4	24	26	28
July	1500	260	2·8	25·5	27	29	32
July	1700	474	3·3	24·6	28	30	32*
August	0930	710	4·9	17·8	24	26	28

7.31 Identify in the table the occasions that give the highest indoor air temperatures (those marked *) and compare the temperatures with those recommended as the maximum allowable for indoor comfort. The *IHVE guide* (page A1-7) recommends that the resultant temperature should not exceed 25°C for summer comfort. *IHVE guide* figure A1.3 shows a relation between resultant temperatures and air temperatures and, for this initial assessment of thermal comfort, it is sufficient to accept as a criterion that an air temperature of 27°C should not frequently be exceeded. All the occasions marked * exceed this criterion by substantial amounts and it is obvious that clear glass will not provide summer comfort in this case.

7.32 *Decision: clear glass is not satisfactory.*

7.33 Examine the value of using a heat-absorbing glass. The shading coefficients of body-tinted heat-absorbing glasses change with thickness much more than those of clear glass. The thickness is therefore a more important factor when heat-absorbing glasses are used and it is necessary, without going into the details of mechanical design which can be left until later, to estimate the order of thickness that will be needed to give adequate resistance to wind loading. Figure 2 in Chapter 1, Function 4 shows that for an area of 7·2 m² (3600 mm × 2000 mm) 6 mm glass with an aspect ratio of 1·75:1 will withstand windloading of no more than 750 N/m². This loading is sure to be exceeded at the site, so glass thickness will have to be at least 10 mm—the next available substance—and it is reasonable to continue the thermal design on the basis of 10 mm glass thickness.

From manufacturers' literature, find the shading coefficients of 10 mm bronze body-tinted heat-absorbing glass ($SWSC = 0{\cdot}31, LWSC = 0{\cdot}24$) and the same glass in double glazing with 6 mm clear glass as the inner leaf ($SWSC = 0{\cdot}25$, $LWSC = 0{\cdot}15$). Recalculate the five key occasions, marked * in the foregoing tables, with heat-absorbing glazing.

Table XIII Recalculated temperature conditions

L	Facade	Month	Time	Gain	Glass	dT	T_0	T_1	T_2	T_3
5 m	SE	July	0915	570	double	7·2	19·4	27	29	31
	NW	July	1700	420	double	5·3	24·6	30	32	34
5 m	SE	July	0930	570	single	3·5	19·9	23	25	28
	NW	July	1700	420	single	2·6	24·6	27	29	32
20 m	SE + NW	July	1700	474	single	2·2	24·6	27	29	31

7.34 Conditions in the private (5 m) offices are probably acceptable if the windows face south-east, but not if they face north-west. Because each private office will have only a single occupant (resultant temperature rise about half a degree) and a negligible amount of machinery and because daylighting will be adequate at times of high indoor temperature, the important value of indoor air temperature is T_1. For the south-east facade, $T_1 = 27°C$ is acceptable, but for the north-west facade, $T_1 = 30°C$ is not.

7.35 In the general offices (15 m and 20 m) the internal heat gains will be as calculated and the operative temperature is T_3—especially as, with heat-absorbing glazing, artificial lighting will often be needed at places remote from the window. None of the examples is acceptable. The borderline value, 28°C, for the 15 m office is eliminated if the private offices must be on the south-east facade. The 15 m office facing north-west is marginally worse than the 20 m office and in practice would have a higher lighting load because daylight is available from only one side.

7.36 *Decision: The private offices will be distributed on the three floors and will face south-east.*

general office
lavs
lifts
stairs
private offices

7.37 The methods used to calculate approximate internal temperatures have probably given values that are somewhat too high, because such factors as the heat loss through the facades and the intermittent use of lights have not been taken into account. They have, however, been useful in whittling the enormous number of originally available window designs to a few possible solutions that can now be examined in detail. The detailed calculation of environmental temperatures (the best basis for design of non-air-conditioned buildings) takes in all relevant aspects of the thermal balance of the building and is described in the *IHVE guide*. It is a lengthy process and most advisory services have devised computer programs that can work through it in a short time so that it is feasible to examine a few alternative designs. The remainder of the thermal design assumes that access to such a program is available.

Calculation of environmental temperatures

7.38 Use the computer to calculate environmental temperature in the general offices with two types of glazing: scheme 1 10 mm heat-absorbing, bronze-tinted, single glazed; scheme 2 10 mm bronze, plus 6 mm clear, double glazed (Table XIV).

7.39 Compare these temperatures with the 27°C recommended as a maximum in the *IHVE guide* (page A8-2). Both schemes give environmental temperatures that are a little too high and depend principally on the internal heat gains. At the hottest times it will probably be sufficient to provide artificial lighting only along a strip 10 m wide, centrally between the windows. This would halve the lighting load and the resulting economy would justify any extra expenditure on switch gear. Recalculate the environmental temperature on this basis (Table XV).

7.40 With the revised lighting schedule, the environmental temperature will be acceptable with scheme 2, double glazing. Check that this scheme is acceptable for the private offices (Table XVI).

Table XIV Calculated environmental temperatures for general office in July, assuming 10 air changes/hour

Time	Environmental temperatures (°C)	
	Single glazing	Double glazing
0900	26·5	26·1
1000	27·5	27·0
1100	28·2	27·7
1200	28·7	28·2
1300	28·9	28·5
1400	28·8	28·5
1500	28·6	28·3
1600	28·8	28·5
1700	28·8	28·4

Table XV Calculated environmental temperature for general office in July, assuming limited lighting and 10 air changes/hour

Time	Environmental temperatures (°C)	
	Single glazing	Double glazing
0900	25·2	24·7
1000	26·2	25·6
1100	26·9	26·3
1200	27·4	26·9
1300	27·6	27·1
1400	27·5	27·1
1500	27·2	26·9
1600	27·5	27·1
1700	27·5	27·0

Table XVI Calculated environmental temperature for private office in July, assuming 10 air changes/hour

Time	Environmental temperatures (°C)
	Double glazing
0900	24·1
1000	24·9
1100	25·5
1200	25·6
1300	25·5
1400	25·0
1500	24·3
1600	24·2
1700	23·9

7.41 *Decision: All windows will be double glazed with 10 mm body-tinted heat-absorbing bronze glass and 6 mm clear glass.*

Conclusion

We have shown how, in buildings and site conditions where very exacting criteria have to be met, designers can systematically check provisionally selected glass types, thicknesses and configurations, for functional suitability. This worked example has considered the proposed glass facade of the building described in para **1.02**, page 31, in terms of the following functions:

- As a *light* filter (view out; task illumination; and sunlight penetration);
- As a *sound* filter (noise insulation); and
- As a *heat* filter (solar heat gain).

Two further checks may be considered necessary. First, ability of the chosen glass to resist wind loading; and second, ability to resist the thermal stresses developed by differential expansion when the sun shines on the exposed parts of the glass while the edges are kept cool by the frames.

The method of checking glass thickness for ability to withstand *wind loading* has already been demonstrated in Chapter 1, Function 4, page 6. Checking for *thermal stresses* is a matter of technical detail which most architects would leave to the technical advisory service of the glazing manufacturer, having satisfied themselves about the basic suitability of the window glass in terms of light, view, sound and heat. The method is described in Chapter 9.

Effect of detailed decisions on original design

The decisions made in the later stages of this worked example have changed some of the original assumptions on which the earlier calculations had been based. It would, in practice, therefore be necessary to repeat some of the stages of design until an acceptable compromise of the conflicting requirements had been found. The recalculations that would be necessary will not be done here—it would be a matter of checking the effect of the bronze-plus-clear double glazing finally adopted, on *view out* (the original assumption was clear glazing, it will be remembered), and so on.

But it should be noted that comfort has been considered only for the summer condition; winter comfort too would need to be considered, and this is discussed in Chapter 1, Function 7, paragraphs 7.41 and 7.42.

Neither conformity with fire regulations nor privacy considerations for the toilet blocks have been discussed and, though the whole of the design has been pursued through technical arguments, the relative costs of the various design solutions will have an important and sometimes an overriding influence. The relevant cost may be the capital outlay but in the present climate of concern for energy conservation it is likely to be the overall running cost of the completed building.

It must be stressed that the design solution postulated here relates only to the entirely hypothetical example that was chosen and might even have been changed if the design had been fully worked out. Many of the decisions that have been imputed to the architect would in reality depend upon his personal preferences and each would influence the outcome of the design. Setting out the design process in the fashion of this worked example makes it seem a laborious process. It is not so in practice and many of the stages that take a lot of space to describe are in fact accomplished by the designer in what appears to be an intuitive manner but is really the rapid application of his experience.

Also, an exercise as detailed as this would normally only be necessary where the functional requirements were highly exacting (eg highly glazed office building in a noisy environment), and it is not suggested that calculations of this complexity are necessary for selecting a suitable window glass in, say, a dwelling with moderately sized windows in normal surroundings.

Part I Window Design

3 Computer programs for environmental design

1.01 Computers offer the means of predicting quickly the outcome of the complex interrelations of environmental design, especially with regard to energy conservation. There are two basic methods of computer operation: batch operation and interactive operation.

1.02 In *batch operation*, the computer takes in a block of data, performs the calculation, prints the results and then stops, leaving itself free for the next user. Data must be presented in a strictly formal order, usually prepared on paper tape or punched cards from previously completed data forms, and it may take a day or more to get the results. Batch operation[1,2] is most suited to the accurate and detailed calculations of environmental factors that are needed towards the end of the design process when most of the design decisions have already been made. It is less useful in the early, exploratory stages of design and it is here that interactive operation is most appropriate.

1.03 In *interactive operation* the user can 'converse' with the computer by means of a typewriter-style keyboard, a visual display unit and light pen, or a similar device.[3] Once the program has been initiated the orderly input of the data is organised by the computer in a question and answer routine in which the user is asked to supply the item next required. The programs described here use English words in preference to codes, are tolerant of mis-spellings, accept curtailed lines of data if the remaining items would be irrelevant and guide the operator when any special format of input is required. The printing is arranged so that, as shown in the example later, the input and output statements fit together to form an easily readable record of the steps in the calculation.

1.04 An important aspect of the available consultancy service is that a portable terminal can be used in the architect's own office where all the information about the design is assembled. The terminal is connected to the computing organisation's network by means of a local call through the normal telephone service. With this method of operation the architect can check the practical feasibility of his ideas before they are established in drawings.

2 Details of calculation process

General principles

2.01 The programs operate by calculating, for each hour of day, the heating or cooling requirement, or the temperature without heating or cooling, by the method of part A of the *IHVE Guide*.[3] If the user is interested in the performance of the building under specific conditions this result is given as output—either for each hour of day or selecting the peak value as required. If energy consumption is required the individual heating and cooling requirements are summed to give a year-round total. In the following account the methods of calculation with and without air conditioning are described together and the differences between them are noted where appropriate.

3 Calculation of hourly heating or cooling requirements

General method

3.01 The method follows closely the method of the *IHVE Guide* with a simplification in the calculation of ventilation heating/cooling requirement.

Solar intensities

3.02 Base values by mathematical formula to give values corresponding to Tables A6.24 to A6.36, *IHVE Guide*.

3.03 Actual values derived by applying coefficients from pA6-6, *IHVE Guide*, to the base values derived as above.

Fabric gains

3.04 Calculated exactly according to section A6 and A8, *IHVE Guide*.

Solar gains through windows

3.05 The input data is in the form of shading coefficients divided into components corresponding to directly and indirectly transmitted radiation. These coefficients are converted within the program to solar gain factors by the equations:

$$S = 0{\cdot}8 \times SC$$

$$Sa = 0{\cdot}8 \times (SCSW \times f + SCLW)$$

where S = Solar gain factor
Sa = Alternating solar gain factor
SC = Shading coefficient
SCSW = Component of shading coefficient corresponding to directly transmitted radiation (shading coefficient—short wave)
SCLW = Component of shading coefficient corresponding to indirectly transmitted radiation (shading coefficient—long wave)
f = Surface factor (*IHVE Guide* Table A8.6).

f takes a value varying from 0·8 for lightweight buildings to 0·5 for heavyweight buildings, where the weight of the building is determined from the average admittance of the room surface.

External obstructions

3.06 These are assumed to intercept only the direct rays of the sun and not to affect the sky or ground diffuse component. The obstruction is specified by its angular height above the horizon at each of the eight major points of the compass as seen from the building. The obstruction's height between these points is determined by linear interpolation.

Internal gains

3.07 Treated as in section A8, *IHVE Guide* for buildings without heating or cooling. For buildings with heating or cooling no account is taken of the storage of internally generated heat in the fabric.

Ventilation gains and houses

3.08 In a building without air conditioning the ventilation coefficient is used as in section A8, *IHVE Guide*.

3.09 With air conditioning the sensible heat gain or loss is calculated as the product of rate of air supply, specific heat of air, difference between outside air temperature and inside environmental temperature.

3.10 Latent heat gain or loss in buildings with air conditioning is calculated from the principles of the 1965 *IHVE Guide*,[3] making the assumption that the absolute humidity of outside air stays constant throughout the day.

4 Calculation of energy used in refrigeration and heating

Basic principle

4.01 The year is divided into its constituent months each of which is assumed to consist of a number of identical days with weather corresponding to the average for the month. On each day the solar intensity, temperatures and humidity vary in the accepted diurnal fashion. The energy consumption for a day is multiplied by 30½ to give the monthly energy consumption.

Climatic data

4.02 Input is for the appropriate data for extreme months of the year. Values for intermediate months are determined by fitting sinusoidal curves between the values for the extreme months.

Calculation of daily consumption

4.03 In a heated building the heating requirements for each hour of day are accumulated to form a total. In an air conditioned building the heating and refrigeration requirements are accumulated separately to form separate totals.

Solar intensities

4.04 The intensities calculated in paragraph **3.03** are modified to provide the radiation to be expected with average cloud cover. This is done by using separate multiplying factors for direct and indirect radiation, derived from an extension to the analysis by Wilberforce of the correlation between total measured radiation, theoretical total radiation and bright sunshine records.[4]

Treatment of fortuitous gains

4.05 In a building without air conditioning all fortuitous gains, ie internal gains and solar gain through the windows, are assumed to be available to offset whatever losses occur at the same time.

4.06 In a building with air conditioning two schedules of operation are represented.

4.07 In general a model representing a two-pipe induction system is employed. Any fraction of the internal gains can be specified to be available to offset losses occurring at the same time. Solar gains are considered to form a cooling load and not to be available to offset losses.

4.08 Also available is a model to represent a variable air volume

Table I Data required for energy consumption calculation in a building without air conditioning.

Data required	Source of data, where reference to published technical information is required	Notes
(1) **INTERNAL CONDITIONS** Controlled temperature (For energy consumption calculation only)		Values greatly different from 20°C are not common in Great Britain. The general question of comfort is discussed in the *IHVE Guide*, section A1.[3]
(2) **ROOM SHELL** (a) Dimensions (b) Basic thermal properties of surfaces (i) Admittance (ii) U Value	For common structures use Milbank and Harrington-Lynn, 'Thermal Response and the Admittance Procedure', BSE, 1974, p.84.[5] *IHVE Guide*, section A8 for item b(i) section A3 for item b(ii), gives more general values.	For more precise calculation of solar gains through opaque fabric the program will also accept supplementary data: (1) Decrement factor and time lag (Milbank and Harrington-Lynn[5] or *IHVE Guide*, section A8). (2) Absorptance of surface (Light, Medium or Dark). (3) Orientation of surface.
(3) **WINDOW APERTURE** (a) Height and width (b) Orientation (c) Inclination (d) No. of windows (e) External shading (i) Depth of shading at top (ii) Depth of shading at side.		The example defines an array of windows; that array could equally well be a single window. Up to three arrays of windows can be specified as existing in a given wall.
(4) **GLAZING** (a) Shading coefficients (directly and indirectly transmitted components). (b) U values.	Pilkington Brothers, *Thermal transmission of windows*.[6]	The directly and indirectly transmitted components are generally referred to as short wave and long wave respectively (hence the abbreviations SCSW and SCLW). This follows the characteristic wavelength associated with directly and indirectly transmitted radiation.
(5) **INTERNAL GAINS** (a) Magnitude of gain (per m^2 of floor) (b) Time when gain is present.	*IHVE Guide*, section A7.	
(6) **VENTILATION** Air changes per hour.	*IHVE Guide*, section A8 when summer overheating is considered.	Values for the air change rate in buildings without forced ventilation are difficult to give with any certainty. Any values used would be estimates rather than firm values. In an air-conditioned building, on the other hand, it could be readily determined or specified.
(7) **CLIMATE** (a) Average January and July values for: (i) Daily average dry bulb temperature. (ii) Daily maximum dry bulb temperature. (b) Months with highest and lowest average per cent of possible sunshine. (c) Per cent of possible sunshine in lowest months.	Meteorological Office tables of climatic data (Values for a large number of locations throughout the world).[7]	The program interpolates between maximum and minimum values to define a typical year. For response to specified conditions give items a (i & ii) for the particular conditions being considered. *IHVE Guide*, sections A2 and A8, gives a comprehensive list of typical design data.
(8) **PLANT DETAILS** (a) Cost of heating fuel. (b) Efficiency of boiler.	(a) Use current tariff data; Conversion factors are: 1p per therm = 0·0095p per MJ 1p per kWh = 0·28p per MJ. Also see *IHVE Guide*, section B18. (b) *IHVE Guide*, section B18.	

system without terminal re-heat—all fortuitous gains are assumed to be available to offset losses.

4.09 In either of the above cases any fortuitous gains not balanced by losses are regarded as a cooling load.

Representation of intermittent operation

4.10 *Weekend shutdown* This is not represented in the program. In calculating the energy required in a building occupied for a five-day week, the program's results would be multiplied by $(5\frac{1}{2} \div 7)$ to allow for five-day operation including the extra energy required to start up after a weekend.

4.11 *Night-time shutdown* The internal temperature is set 5°C lower than its daytime value when the building is unoccupied to allow for the effect of switching off the heating. Refrigeration (where considered) is taken to be in operation from the time the building is vacated until midnight. Any refrigeration considered by the program to be required between midnight and start-up time would correspond approximately to pre-cooling requirements.

5 Example of the program in use

5.01 **1** and **2** show a typical office building which might be considered from the point of view of energy consumption and susceptibility to summer overheating, assuming that there is no

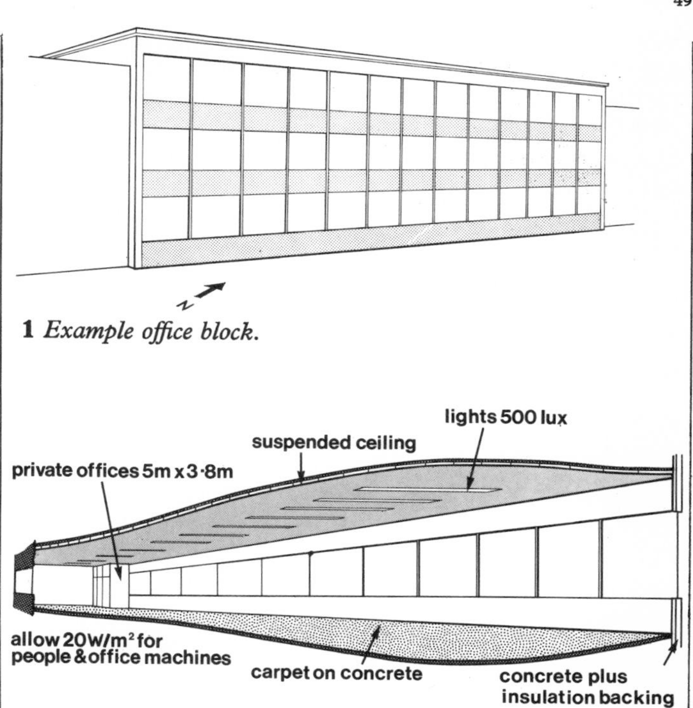

1 *Example office block.*

2 *Section through first floor.*

```
                          PAGE 1

CALCULATION STARTED  1354  25 JUNE 1976

 SPACE NOT AIR-CONDITIONED
 PROGRAM LAST REVISED 20 JANUARY 1976


 JOB NO?EXAMPLE

*TEMPERATURES REQUIRED, ENVIRONMENTAL OR AIR?ENVIRONMENTAL
```

Notes (not part of the print-out)

Request the program to consider the environmental temperature (which is a weighted mean of air and mean radiant temperature) as recommended by the IHVE.[3]

```
*ROOM DIMENSIONS
*
*LENGTH?50
*WIDTH? 20
*HEIGHT? 3
```

Consider a whole floor for energy consumption calculation.

```
*ROOM SURFACE DESCRIPTION
*
*TABLE
*                           /-> FOR SUN ON FABRIC
*SURF   ADMITT-   U VALUE   DEC'NT    TIME   ABSORP-   AZTH   INCLN
*        ANCE     IF EXT'L  FACTOR    LAG     TANCE          (IF NOT 90)
*   1?    1        0.8
*   2?    1.2
*   3?    1        0.8
*   4?    1.2
*CEILING? 2.1
*FLOOR    2.8
```

Surfaces 1 to 4 are walls numbered clockwise starting with a long wall.
Walls 1 and 3 are external and so are specified by both admittance and U-values.
Walls 2 and 4, and also the floor and ceiling, make negligible contribution to the conduction from the space and so are specified simply by their capacity to store solar heat, i.e. the admittance.

```
*WINDOWS
*
*TABLE
*SURF   WDW     HT    WTH    AZTH    INCLN    NO IF M'PLE
* NO    NAME
*? 1    MAIN    2     3.8    135      90         13
*? 3    MAIN    2     3.8    315      90         13
*  (carriage return)
*
*
*ANY SHADING?NO
```

Let wall no. 1 be the south-east wall then the windows in it also face south-east.
Similarly the windows in surface 3 face north-west.
Thirteen windows, each 2 m high by 3·8 m wide, are specified in each of these walls.
Note that each array of windows is given a name whose main purpose is to identify it if more than one array is specified for a given wall.
If the user answers a question by a blank line (i.e. a carriage return with no input preceding it) this means that he has put in all his data under this heading and the program moves on to the next item.

PAGE 2

```
*GLASS AND BLIND PROPERTIES
*
*NO OF SCHEMES?3
*
*TABLE
*SCHEME     SCSW    SCLW    U VALUE
*   1?      0.92    0.05      5.6
*   2?      0.74    0.10      3.0
*   3?      0.41    0.14      3.0
```

Three types of glazing are considered
Scheme 1 is single clear glass
Scheme 2 is double clear glass
Scheme 3 is double glazing of heat-absorbing glass and clear glass.

```
*INTERNAL GAINS
*GIVE GAINS PER SQUARE METRE
*
*NO OF SOURCES?2
*
*TABLE
*       SIZE     START    END
*      (WATTS)   TIME     TIME
*NO 1?  25        900     1800
*NO 2?  20        900     1700
```

No 1 is lighting, 500 lux from fluorescent tubes in diffusing fittings.
No 2 is occupants and sundry office machinery.

```
*VENTILATION
*
*AIR CHANGES PER HOUR?2
```

Estimate two air changes per hour during the time when heating is used.

```
*OUTSIDE CONDITIONS
*
*WHICH MONTH?YEAR
*
*INTERNAL DESIGN CONDITIONS
*
*TEMPERATURE?20

*OUTSIDE CONDITIONS
*
*AVERAGE WEATHER
*                 TEMPERATURES
*            DAILY MEAN   DAILY MAX
*JANUARY?       4.3          6.3
*JULY?         17.7         21.8
*
*
*PERCENT OF POSSIBLE SUNSHINE (CAMPBELL-STOKES RECORDING)
*HIGHEST MONTH?JUNE
*PERCENT SUN?43
*LOWEST MONTH?DECEMBER
*PERCENT SUN?16
*
*
```

When the computer asks for the month for calculation, a reply of 'year' indicates that year round performance (ie energy consumption) is required. Having been told that a heating energy consumption calculation is wanted, the program asks for the temperature to be maintained during the heating season. The values given for outside temperature and sunshine are taken from meteorological records. Kew is the example here but data exists for a large number of locations.

```
*ENERGY COSTS AND PLANT DATA
*
*COST OF HEATING FUEL (PENCE PER MEGAJOULE)?0.11
*EFFICIENCY OF HEATING PLANT (PERCENT)?70
```

Energy costs are based on a gas tariff of 11·5p per therm. The heating plant efficiency corresponds to continuously run central heating with radiators.

air conditioning. The same building was used in the worked example in Chapter 2. Table I lists the data required and the computer document is shown on pages 49-53. The sequence of the computer calculation is:

1 Input basic data to determine energy consumption and initial results.
2 Determine whether a reduction in the ventilation rate will save a significant amount of energy.
3 Test for summer overheating in a selected area.
4 Specify a remedy for overheating and test its effectiveness.

5.02 This is a shortened version of what would happen in practice where the designer would want to consider daylight (particularly with the window area as reduced by step 4) and the modified energy consumption with the revised windows. However, it illustrates the process (initial definition, performance appraisal, design modification and reappraisal) as aided by the computer program.

5.03 Because the program is conversational in operation, the usual distinction between the input document and the print-out does not apply and both input and output appear on the same sheet of paper. In this example the information typed by the user has been underlined to distinguish it from the computer's output.

PAGE 3

```
*CASE NAME?VALUES FOR A WHOLE FLOOR 50M X 20M
```

```
RESULTS VALUES FOR A WHOLE FLOOR 50M X 20M

ANNUAL FUEL CONSUMPTION (MEGAJOULES) AND RUNNING COSTS (POUNDS)

           ENERGY     COST
SCHEME 1   400524     441
SCHEME 2   328419     361
SCHEME 3   342011     376
```

These results give the consumption with each of the glasses specified, eg Scheme 1 is single clear glass.
Conclusions—double glazing would save about £80 pa and using heat-absorbing glass would cost £15 pa in energy more than double clear glass.

```
JOB FILE UPDATED WITH CURRENT DATA

CHECK VALUES?NO

*MODIFICATIONS?VENTILATION
*
*AIR CHANGES PER HOUR?1.5


*MORE CHANGES?NO
*
*CASE NAME?SAME
```

In the printout 'CHECK VALUES ?' the program is offering an analysis of some intermediate steps in calculation.

Consider the possible savings if the ventilation is reduced to a minimum level.

```
RESULTS VALUES FOR A WHOLE FLOOR 50M X 20M

ANNUAL FUEL CONSUMPTION (MEGAJOULES) AND RUNNING COSTS (POUNDS)

           ENERGY     COST
SCHEME 1   340611     375
SCHEME 2   269209     296
SCHEME 3   282378     311
```

Conclusion—reducing the ventilation will save approximately £65 pa in heating energy.

```
JOB FILE UPDATED WITH CURRENT DATA

CHECK VALUES?NO
*MODIFICATIONS?DAILY DESIGN
```

Ask for calculation of response to specified design conditions.

PAGE 4

```
*DAILY DESIGN
*
*DATA USED FOR JULY
*DAILY AV. TEMP. TAKEN AS.       19.00
*DAILY MAX. TEMP. TAKEN AS       25.50
```

The program supplies meteorological data for Kew observatory for design conditions in July—other months could be selected by the user.

```
*MORE CHANGES?ROOM
*
*LENGTH? 5
*WIDTH?3.85
*HEIGHT? 3
```

Consider the small private offices, since they will be most susceptible to overheating. Note that the length (ie the longest dimension) is the depth from the facade to the back of the room.

```
*MORE CHANGES?MATERIAL
*ANSWER INTERPRETED AS ROOM MATERIALS
*
*TABLE
*                                /-> FOR SUN ON FABRIC
*SURF   ADMITT-   U VALUE   DEC'NT   TIME   ABSORP-    AZTH    INCLN
*        ANCE     IF EXT'L  FACTOR   LAG    TANCE              (IF NOT 90)
*? 1     1.2
*? 2     1           0.8
*? 3     1.2
*?(carriage return)
```

The program will still consider surfaces 1 and 3 as the external walls since this is how they were originally defined. Now that a short wall is an external wall it is necessary to re-specify one of the short walls (in this case number 2) as external. Surfaces 1 and 3 now become internal walls and must also be re-defined.

```
*MORE CHANGES?WINDOW
*
*TABLE
*SURF   WDW   HT   WTH   AZTH   INCLN   NO IF M'PLE
*.NO    NAME
*? 1  DEL
*ALL GLAZING DELETED IN SURFACE 1
*? 3  DEL
*ALL GLAZING DELETED IN SURFACE 3
*? 2    MAIN  2   3.8   135   90
*?(carriage return)
```

Since surfaces 1 and 3 are now internal walls delete the windows in them. Specify the window in the external wall, surface 2, leaving the column 'NO IF M'PLE' blank as the program assumes just one window.

```
*MORE CHANGES?VENTILATION
*
*AIR CHANGES PER HOUR?10
```

Assume that extensive areas of open window will be available to provide increased ventilation in peak conditions.

```
                              PAGE 5

*MORE CHANGES?NO
*
*CASE NAME?SMALL PRIVATE OFFICE ON SE FACE :

______________________________________________

RESULTS SMALL PRIVATE OFFICE ON SE FACE
        JULY     10 AIR CHANGES/HOUR


PEAK ENVIRONMENTAL TEMPERATURES

SCHEME 1       36.9 AT 1100
SCHEME 2       36.7 AT 1100
SCHEME 3       32.6 AT 1200

______________________________________________

JOB FILE UPDATED WITH CURRENT DATA

CHECK VALUES?NO


*MODIFICATIONS?WINDOW
*
*TABLE
*SURF    WDW   HT   WTH    AZTH    INCLN      NO IF M'PLE
* NO     NAME
*? 2     MAIN  1.8  0.5    135      90             2
*?(carriage return)


*MORE CHANGES?NO
*
*CASE NAME?REDUCED WINDOW AREA

______________________________________________

RESULTS REDUCED WINDOW AREA
        JULY     10 AIR CHANGES/HOUR

PEAK ENVIRONMENTAL TEMPERATURES

SCHEME 1       28.4 AT  1300
SCHEME 2       28.1 AT  1300
SCHEME 3       27.3 AT  1300

______________________________________________

JOB FILE UPDATED WITH CURRENT DATA

CHECK VALUES?NO

*MODIFICATIONS?NO
*
DATA STORED FOR RE-USE
```

Temperatures are far too high for comfort; Drastic modifications will be required.

Reduce the window area to two vision slits each 1·8 m high by 0·5 m wide.

Temperature will be reasonably comfortable. The incidence of overheating will be slightly reduced by the use of heat-absorbing glass.

References

1 *Air conditioning loads by computer*. Environmental Advisory Service Report. Second edition. Pilkington Brothers Limited. April 1974.
2 *Obstructional shading by computer*. Environmental Advisory Service Report. Pilkington Brothers Limited. May 1973.
3 Institute of Heating and Ventilating Engineers. 1970 *Guide*. Published 1971, reprinted 1975.
4 Wilberforce, R. R. *The effect of solar radiation on window energy balance*. International CIB Symposium, Energy Conservation in the Built Environment, 6-8 April 1976.
5 Milbank, N. O. and Harrington-Lynn, J. 'Thermal response and the admittance procedure'. *Building Services Engineer*, 1974, **42** (May).
6 *Thermal transmission of windows*. Environmental Advisory Service Report. Pilkington Brothers Limited. March 1973. Third edition.
7 Meteorological Office. *Tables of temperature, relative humidity, precipitation and sunshine for the world*. Met. O. 856, London HMSO.

Part I Window Design

4 The flow of light in lighting design

Introduction

Codes of practice on lighting generally distinguish between lighting a task for efficient performance and lighting a room for pleasant appearance. As the contribution to efficient performance made by an environment that is comfortable and pleasant —visually, thermally and acoustically—has become more widely recognised the designer has been encouraged, especially in the design of modern open-plan offices, to revise his attitude to the separate but interacting roles of daylight and artificial light.

That light gives the impression of flowing into a space is a useful idea that has gained steadily in acceptance since it was expressed by Lynes and others in 1966.[1] This chapter discusses the concept from the point of view of the designer and shows how natural and artificial light can be engineered in ways that get closer to those aspects of the visual environment that we sense and subjectively feel to be important, but which previously have been beyond the scope of engineering technology.

Background

The first people to measure lighting were the scientists and engineers who had made the early electric lamps. The lumens from these sources were precious and the lamps were suspended low over the work-bench with conical shades to intercept the upward light and redirect it downwards. To assess the effectiveness of their inventions they developed a photoelectric cell consisting of a small flat disc of light-sensitive material that was placed on the working surface to enable the density of lumens to be measured and expressed as lumens per unit area. The density of lumens is now called illuminance and the unit is the lux.

To specify the average illuminance from an artificial lighting installation seemed quite appropriate, but for daylight illumination inside buildings a more elaborate concept was required. Because the illuminance out of doors is continually changing so too must the indoor daylight illuminance change. The indoor illuminance was, therefore, expressed as a percentage of the outdoor value: this is the daylight factor.

When illuminance measurements moved out of the workshop it was not always easy to say which was the working surface so the rule became that, when in doubt, a horizontal plane 0·85 m above floor level should be used as the reference plane for specifying and measuring illuminance. In this way the concept of a horizontal working plane came to be applied not only to workshops and offices (working situations) but also to reception rooms, foyers, auditoria, stairways, corridors (non-working situations) and many other locations for which the average illuminance on this imaginary horizontal plane appears to have no particular relation to the lighting of the things that people look at. Only recently have the various codes and recommendations for lighting in buildings acknowledged the distinction that must be made between the requirements for working and non-working situations. In working situations the principal visual tasks can be specified and the light must be distributed throughout the space to enable the functional visual requirement to be satisfied at any working location. In non-working situations the effectiveness of the lighting cannot be assessed in terms of the standard of task visibility or the degree of visual comfort that it provides. The main problem for the code maker is that there are very few rules that lead to universal success in the lighting of these non-working situations.

One rule that does emerge time after time is that there must be differences of illumination within the space: in other words, the uniformity that is generally recommended to allow organisational flexibility in working situations should be deliberately avoided where lighting is to be used creatively to contribute to the quality of the visual environment. Although this is easy to recognise, it has not been easy to prescribe because the traditional terms and units of illuminating engineering fail to identify the aspects of lighting that are relevant to this purpose.

How much light is enough?

It is recommended[2,3] for classrooms with side windows that nowhere shall the daylight factor be less than 2 per cent, and for factories with rooflights, 5 per cent. Does this mean that factory workers require two and a half times as much light as school children? There is no corresponding difference between the recommended illuminances for artificial lighting. However, inspection of classrooms and single-storey factories suggests that the recommendations are about right and that if these minimum daylight factors are not achieved artificial lighting is likely to be used throughout the day.

To study the reasons for the discrepancies a scale model of a group of people and working surfaces was arranged so that it could be illuminated by light from various directions; sometimes from overhead, sometimes from the side, sometimes from angles between.[4] For each direction of lighting, observers adjusted the amount of light so that the model scene appeared to be illuminated to the same extent as a standard comparison scene. The illuminance was measured on the horizontal working plane (E_h) and also by two special photoelectric cells one in the shape of a vertical cylinder to record the average illuminance of vertical surfaces (mean cylindrical illuminance, E_c) and one in the shape of a sphere to record the average illuminance of all surfaces regardless of their orientations (mean spherical illuminance, E_s). In every case, the observers' assessment of the strength of illumination of the group of objects was the same but, as shown in **1**, only the mean spherical illuminance maintained an approximately constant value. In particular the experiment made clear the nature of the perplexity of lighting recommendations related to the horizontal working plane: at 90° the curve of horizontal illuminance (E_h) has a value about two and a half times its value at 20°. Thus, when the direction of lighting is approximately 20° above the horizontal, as would be expected at the back of a side-lit classroom, an equivalent assessment of adequacy of the

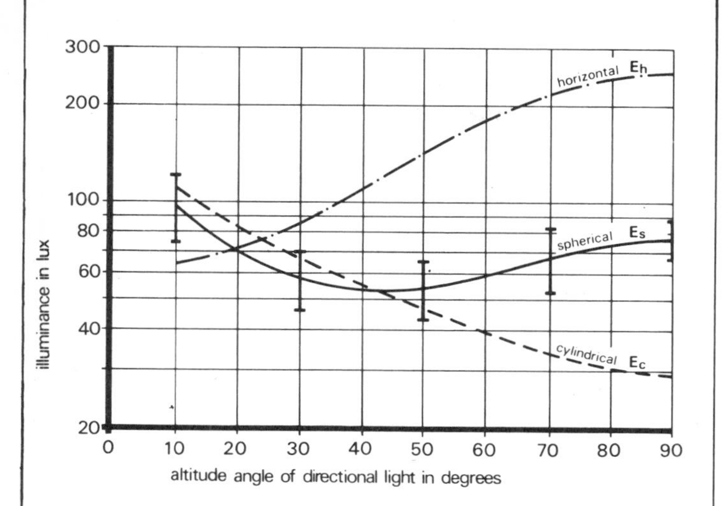

1 *The relations of the three measures of illuminance to the angle of altitude of directional light. The standard deviations about the mean values of* E_s *are shown: the scatter of the results for* E_h *and* E_c *is similar.*

illumination from overhead lighting, as in a single-storey factory, would require an illuminance about two and a half times as great.

For the purpose of specifying illuminance for a situation in which the assessment of the lighting will be based upon the appearance of solid three-dimensional objects, the illuminance upon a horizontal plane is a poor criterion because of the manner in which it depends upon the direction of lighting. Mean spherical illuminance is more satisfactory.

The flow of light

2(a) and **2(b)** show two views of an entrance hall. So great is the change of character that it needs a second look for reassurance that it is really the same room and that the difference is due only to the lighting. As we look more closely at the girl **3(a)** and **3(b)**, it is evident that this change of character is due mainly to the different directional qualities of the two light conditions. 'The flow of light' is, for want of a better term, a convenient way of referring to the subjective impression of strength and direction of lighting within a room. It is not a physical quantity that can be measured with an instrument but, because it is an aspect of man's visual perception of his environment, there are measurable aspects of the physical environment that relate to this perception.

The day and night views show the same objects but with strikingly different patterns of light and shade. These are the physical characteristics of directional lighting.

The apparently complex lighting patterns can be analysed without too much difficulty because they are compounded from three distinct components: *the highlight pattern*, *the shadow pattern* and *the illumination pattern.*

2(a)

2(b)

2 *The change in appearance of an indoor space when daylight from side windows is replaced by artificial lighting from various sources.*

3(a)

3(b)

3 *Changing the flow of light changes the shadow patterns and affects the appearance of an object.*

4(a)

4(b)

4 *A comparatively lifeless portrait* **4b** *is produced by removing the highlights from the eyes.*

The highlight pattern is a pattern of specular reflections of the light source. Even human skin is sufficiently glossy to punctuate the lighting pattern with reflected highlights and the close-up appearance of a face is considerably affected by reflected highlights in the eyes as the re-touched photograph **4(b)** shows.

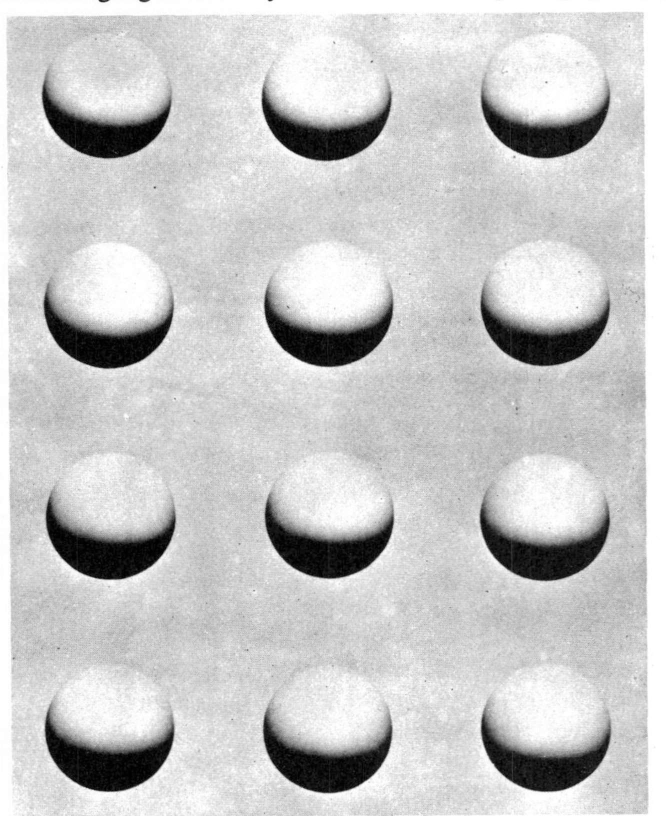

(Photograph reproduced by permission of Philips Electrical Ltd.)

5 *Light incident from an unexpected direction can confuse our perception.*

The shadow pattern is the pattern of shadows that form in the hollows of an object, varying according to the density of the shadows and whether they have sharply defined or blurred edges. A simple pattern of sources can enhance appearance but a muddled arrangement of many sources produces a confusing pattern of conflicting shadows.

The illumination pattern. A solid object in directional lighting produces a pattern caused by the differences in illumination of its surface that is quite distinct from the shadow pattern. A convex surface will show grades of surface illumination not caused by one part casting a shadow on another.

The pictures of the entrance hall show every non-transparent object displaying its own individual and personal lighting patterns which differ not merely in strength and direction but even more in the balances of highlight, shadow and illumination patterns. Lighting that has the capacity to form these patterns clearly and distinctly presents to the observer a rich variety of object appearances as contoured surfaces take on illumination patterns to match their forms, concave and convex surfaces are distinguished by patterns of contained shadow, and glossy surfaces show thin highlights.

Another general rule of lighting is that the type of lighting we like is not that which produces a particular type of object appearance but that which produces variety of object appearance. An understanding of how these lighting patterns are formed is crucial to achieving control over the quality of the visual environment: a detailed description is given in 'Lighting patterns and the flow of light'.[5] The highlight and shadow patterns depend mainly upon the intensity and angular size of the light source. A single, small, high intensity source gives 'sharpness' to the lighting by the nature of the highlight and shadow patterns that it produces. For some objects a high degree of sharpness improves appearance and aids visibility, but this is not always the case. In particular for looking at faces a source of angular size sufficient to give some spread of the highlights and to soften the edges of the shadows is to be preferred. **3(b)** demonstrates what happens if the angular size is too large and sharpness is altogether lost. Highlight and shadow patterns can occur only if an object has sufficient gloss and surface hollows, but any opaque object perceived to intercept a flow of light forms an illumination pattern. It is from the appearance of these patterns that the unifying perceptions of lighting develop. Compact highlight patterns and clean-cut shadow patterns characterise sharpness, while the perception of varying strength and direction of the flow of light relates to the appearances of the illumination patterns on all the objects that surround us. It is not a point of perceptual confusion that matt and glossy surfaced objects generate lighting patterns that are quite different in appearance provided that their appearances are consistent with the overall perceptions of lighting. It has been stated before that the term 'flow of light' describes not a physical quantity but an aspect of our perception. That this concept is fundamental can be demonstrated with **5** which shows hemispheres on a flat surface but if the figure is inverted the solid hemispheres become hollow depressions. Figures have been used in this way to make the point that our perceptions of objects are not developed from independent analyses of their images but perception is a continuous process of making sense of our environment by comparing images both one with another and with our expectations. Experience tells us that a flow of light from above is normal and natural whereas a flow of light from below is reserved for lighting pantomime devils and other unnatural apparitions. Of all the possible perceptions, the system selects that which is most likely. The figure may be inverted but the flow of light holds its direction. With the figure near a window or in some place where there is a lateral flow of light, the reversal effect works with the figure held on its side or even flat on a table. The memory of sky above and earth beneath is not the limit of our understanding of the flow of light for, in any situation, accurate perception of the flow is essential to enable us to make sense of the variety of images presented to our eyes and this develops from appraisal of the illumination patterns on the various surrounding objects.

Scalar and vector quantities of light

New lighting terms and quantities that can take account of appearance are not necessarily more complicated than those already in use. Their strangeness will reduce as familiarity with them grows from their inclusion in the 1973 *IES Code*.[2]
The illumination pattern on an object is produced by the interaction of the three-dimensional form of the object with the spatial distribution of illumination. This distribution is represented by the illumination solid **(6)** for which the distance in any direction from the reference point, O, to the surface of the solid is proportional to the illuminance in that direction. A scale model of the illumination solid at a point would provide a complete, but cumbersome, specification of the illumination at the point. A single value, such as the horizontal illuminance, is more convenient but there is clearly a substantial loss of information. The value of mean spherical illuminance as an index of the quality of the illumination of three-dimensional objects has already been shown and it can now be seen that this is the average value of the illumination solid. It is called scalar illuminance to emphasise that its value—like that of room temperature, for example—can vary from point to point but does not depend upon the orientation of the measuring instrument.

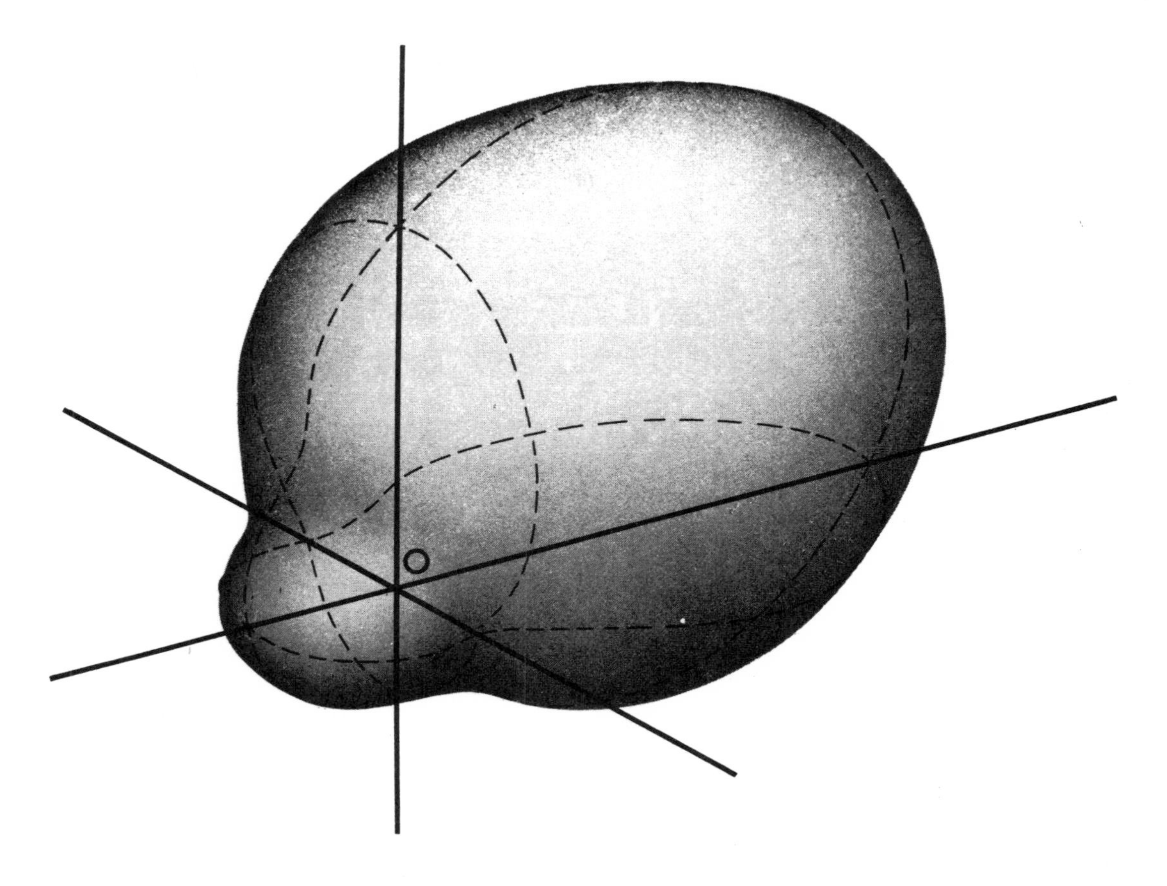

6 *The illumination solid.*

7 shows sections through two illumination solids. In **7(a)** the illumination solid is almost symmetrical about the reference point, O, and the flow of light will appear to be very weak. In **7(b)** the illumination solid is very asymmetrical about O and the flow of light will be strong. One limitation of scalar illuminance is that it does not distinguish between these two situations and there is, therefore, a need for quantities that take direction into consideration. A section of an illumination solid for a reference point, O, is shown in **8(a)**. The illuminance in the direction BO is greater than in the direction AO and a point, C, can be chosen so that OC represents the difference in illuminance (BO-AO). If this process is repeated for all directions, it can be shown[5] that the points, C, C_1, C_2 . . . , lie on the surface of a sphere that passes through O, **8(b)**. This sphere represents the asymmetrical component of the illumination solid and the remaining component must always be a solid that is symmetrical* about O, **8(c)**.

A diameter of the asymmetrical component terminating at O **8(d)** is called the illumination vector (E). Its length is proportional to the maximum difference of illuminance in opposite directions at the reference point and its direction corresponds to that in which the maximum difference occurs. Only one sphere can be described with the illumination vector as a diameter so, when the magnitude and direction of the illumination vector have been specified, the illuminance differences for all other directions are automatically specified and the unique sphere is called the vector solid.

From **7** it was concluded that the strength of the flow of light is related to the degree of asymmetry of the illumination solid. The illumination vector corresponds to the size of the asymmetrical component and the scalar illuminance to the average size of the complete illumination solid. The ratio of these two indicates the degree of asymmetry of the illumination solid so the vector/scalar ratio (E/E_s) is the index of the strength of the flow of light.

* Symmetry may be specified about an axis, a plane or a point. About an axis of symmetry each side appears as the mirror image of the other, and correspondence of any part to its counterpart can be traced along a line normal to the axis. For an object that is symmetrical about a point, correspondence is traced through the point so that any one half of the object appears as the inverse of the other.

Design recommendations

Table I shows how the assessments of a typical observer are likely to relate to various vector/scalar ratios in typical indoor lighting situations. No single value is right for all purposes but the selection of an appropriate value can help to achieve effective lighting, whether its purpose is to enable accurate recognition of a particular object or to give a generally attractive appearance to a variety of objects. There are many situations such as foyers, reception and circulation areas in which the lighting quality is likely to be judged primarily by the manner in which it reveals the human features. To achieve an appropriate strength of flow of light it is generally recommended that the vector/scalar ratio should be within the range from 1·2 to 1·8. For more formal or distant communication, such as a lecture theatre situation, a higher value in the range 1·8 to 2·2 is advisable. It has also been found that a vector direction between 15° and 45° above the horizontal is preferred to a predominantly downward flow of light.

Often it is not practicable to achieve this effect from an artificial lighting installation but, even where artificial lighting is used throughout the working day, the lighting quality can be improved for much of the day by the addition of side lighting from windows. Table II indicates the extent to which the preferred quality of lighting is achieved in a room having windows on one side and a regular artificial lighting installation.

Measuring scalar and vector quantities

Descriptions of how to make instruments for measuring scalar illuminance and the illumination vector have been published elsewhere,[1,4] but it is not necessary to have a special instrument to achieve reasonably reliable results. The simplest method for field measurements is to support a small cube at the reference point with its top face horizontal and its vertical sides parallel to the walls of the room. A conventional photoelectric light meter can be used to measure the illuminance on each face of the cube and the average of the six readings gives the scalar illuminance. For strongly directional lighting this method of measurement can give appreciable error, but the error is unlikely to be significant for measurements of conventional indoor lighting installations.

For each of the three pairs of opposite sides of the cube the difference in illuminance is the component of the illumination

Table I Vector/scalar criteria for lighting design.

Vector/scalar ratio	Strength of the flow of light	Typical situation	Typical appraisal
3·0	Very strong	Selective spotlighting. Direct sunlight.	Strong contrasts: detail in shadow is not discernible.
2·5	Strong	Low BZ*, low FFR†, dark floor. Windows on one side, dark surfaces.	Noticeably strong directional effect: suitable for display but generally too harsh for human features.
2·0	Moderately strong	Low BZ with medium or light floor. Medium or high BZ. Side windows with light surfaces. p.s.a.l.i.	Pleasant appearance of human features for formal or distant communication.
1·5	Moderately weak	Low BZ with medium or light floor. Medium or high BZ. Side windows with light surfaces. p.s.a.l.i.	Pleasant appearance of human features for informal or close communication.
1·0	Weak	Medium or high BZ with light floor. Side windows in opposite walls.	Soft lighting effect for subdued contrasts.
0·5	Very weak	Luminous ceiling or indirect lighting with light surfaces.	Flat shadow-free lighting: directional effect is not discernible.

* British Zonal classification † Flux fraction ratio.

Table II The percentage of normal working hours (0900-1730) throughout the year for which the vector altitude due to both daylighting and artificial lighting will fall below certain angles. [It is assumed that the vector altitude due to daylight alone is not greater than 20° and that the floor is of medium lightness (reflectance, 0·2). The prediction of sky components is described below and in IES Technical Report No. 4.[7]]

	Average illuminance on horizontal plane due to artificial lighting alone											
	200 lux			400 lux			700 lux			1000 lux		
	Maximum vector altitude											
Sky component on horizontal plane	45°	60°	75°	45°	60°	75°	45°	60°	75°	45°	60°	75°
1 per cent	73	88	94	46	75	90	5	56	82	0	39	74
2 per cent	87	93	95	73	88	94	53	78	91	30	68	88
3 per cent	91	94	95	82	92	95	70	86	93	55	79	92
4 per cent	93	95	95	87	93	95	76	89	94	66	84	93
5 per cent	94	95	95	90	94	95	81	91	95	73	88	94

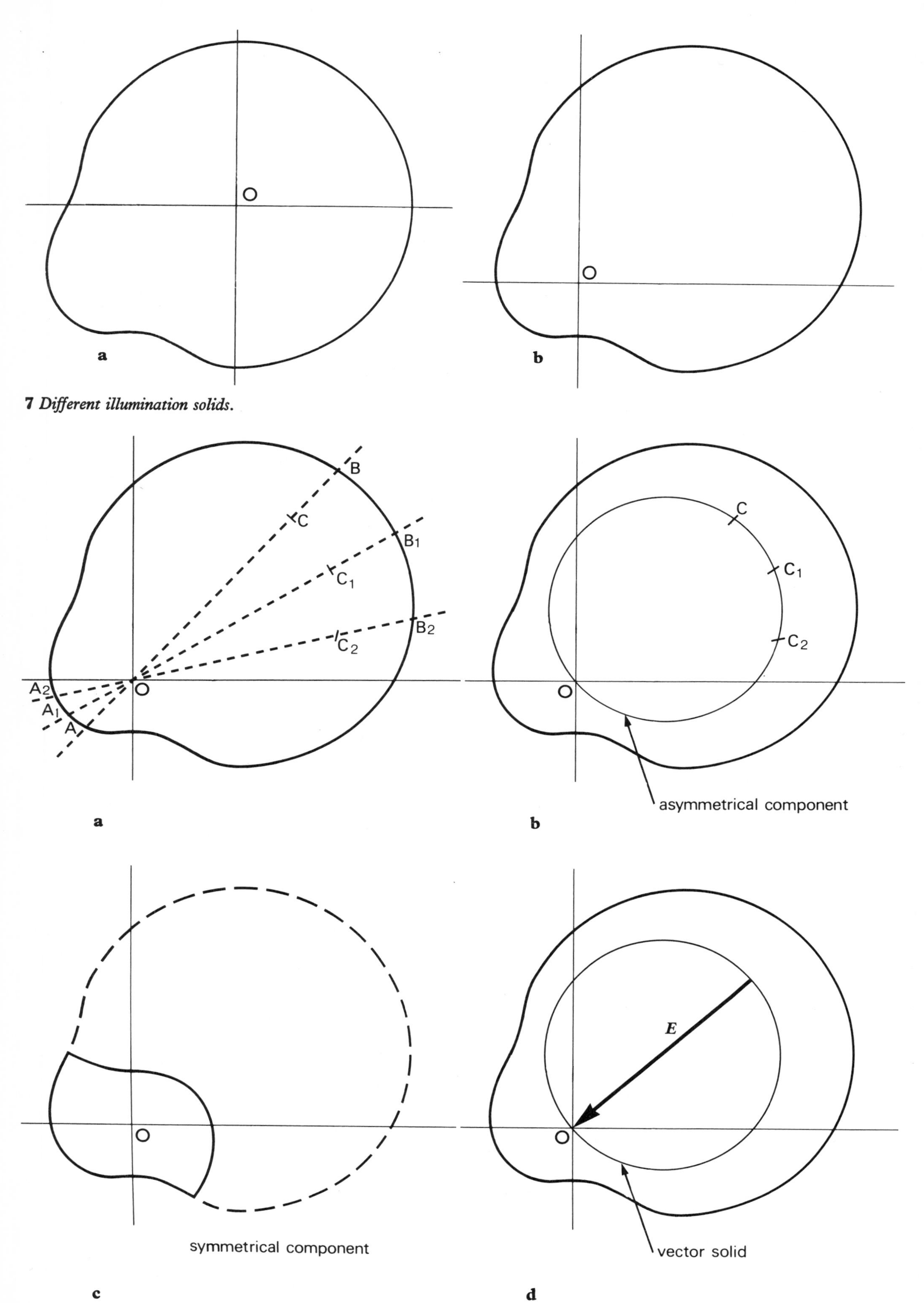

7 *Different illumination solids.*

8 *Symmetrical and asymmetrical components of the illumination solid.*

vector in the direction perpendicular to the side having the greater illuminance. The illumination vector is found by adding the three components vectorially, **9**. Although this method requires only a conventional photoelectric cell the repeated resolution of the illumination vector could be tedious if a large number of readings had to be made. A fairly simple instrument can be constructed to overcome this by cementing half a table-tennis ball over a photoelectric cell and, by trial and error, adding masking until the cell measures the mean hemispherical illuminance.

The cell is mounted, with the minimum of optical obstruction, so that it may be universally rotated to find the direction in which the mean hemispherical illuminance reaches a maximum. The cell is then rotated through 180° to obtain the minimum reading. The sum of the two readings is twice the scalar illuminance and their difference is half the illumination vector.

Calculating scalar and vector quantities

Overhead lighting (regular arrays of lighting fittings or roof-lights that can be classified in the BZ system).

In both the prediction and the measurement of vector/scalar ratios the main problem is that it is not possible to determine the value of the illumination vector until its direction also has been found. However, when lighting is achieved by a regular installation of conventional overhead sources, whether roof-lights or electric luminaires, it can be assumed that the average direction of the illumination vector will be vertically downwards and that, except near walls or other obstructions, the vector/scalar ratio will be approximately constant throughout the space. In this case it is reasonable to specify the lighting in average values and the prediction techniques are simplified.

The illumination vector, E, and the scalar illuminance, E_s, are both proportional to the average horizontal illuminance, E_h, which cancels out in the vector/scalar ratio and does not need to be separately calculated.

$$E = E_h\,(1-\rho_f) \quad (1)$$

$$E_s = E_h\,(K+0{\cdot}5\rho_f) \quad (2)$$

$$E/E_s = (1-\rho_f)/(K+0{\cdot}5\rho_f) \quad (3)$$

Equation 1 arises from the assumption that the vector direction is vertically downwards so that the illuminance in the upward direction is the product of the horizontal illuminance and the reflectance of the floor cavity, ρ_f. The constant, K, used in Equations 2 and 3 is derived from the charts in **10**. From the chart for the appropriate wall reflectance, two values of K can be read: for the downward light K_{BZ} is read from the appropriate British Zonal classification (BZ) curve; for the upward light K_C is read from the C curve. The value of K, to be used in calculating the vector/scalar ratio from Equation 3, is given by

$$K = K_{BZ} + (K_C - K_{BZ}) \times \frac{FFR \times \rho_c}{1 + FFR \times \rho_c} \quad (4)$$

where FFR is the flux fraction ratio and ρ_c is the reflectance of the ceiling. BZ classifications and flux fraction ratios of electric luminaires are usually given by the manufacturer. For typical roof-lights the corresponding BZ classifications are tabulated in Table III: the flux fraction ratio of any roof-light is zero so that the second term of Equation 4 reduces to zero and, for roof-lights, $K = K_{BZ}$.

Values of K that give various values of vector/scalar ratio are shown in **11**, related to the mean reflectance of the floor cavity. Designing for these overhead lighting situations is fairly straightforward. Achievement of an appropriate vector/scalar ratio will ensure the right sort of strength of flow, but the effect of the direction of flow on the appearance of human features will be capable of improvement. To achieve this involves calculating the component vectors and adding them as in **9**.

Light transfer ratios, LTR, are included in Table III so that, when the tabulated BZ classifications have been used to obtain the utilisation factor, UF, from Tables 7·1 to 7·10 of *IES Technical Report No.* 2, the average daylight factor, D, on a horizontal working plane can be calculated from the equation

$$D = 100 \times UF \times LTR \times G \times (g/f)$$

where G is the visible transmittance of the glass, incorporating maintenance factors if necessary.

Side lighting (special luminaires and windows that cannot be classified in the BZ system).

The sky component, SC_h, of the daylight factor can be predicted by a graphic technique using the dotted diagram of **12** for the horizontal plane.[9] Superimpose on the diagram an elevation of the window, drawn to a scale of 1:100, with the origin, O, at the centre of vision and the horizon line horizontal. If the number of diagram dots appearing within the area of clear sky seen through the window is N, the sky component is N/10 per cent at a point 3 m inwards from the window wall. For other distances from the window the scale of the window elevation must be changed: for 6 m inwards the scale should be 1:200, and so on. Dots that appear within the outline of any

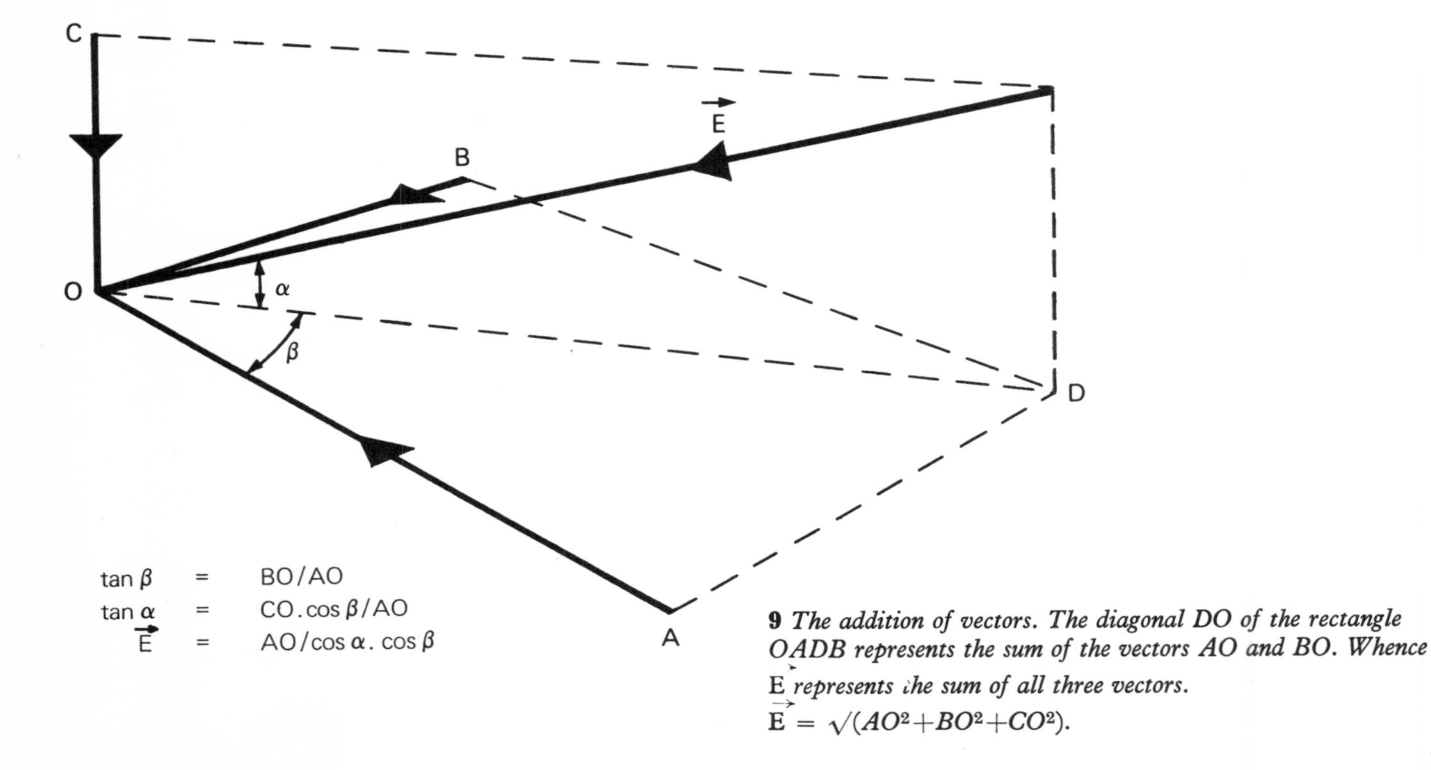

9 *The addition of vectors. The diagonal DO of the rectangle OADB represents the sum of the vectors AO and BO. Whence* $\vec{E}$ *represents the sum of all three vectors.*
$\vec{E} = \sqrt{(AO^2+BO^2+CO^2)}$.

external obstruction seen through the window represent the externally reflected component. Forty per cent of their number multiplied by the average reflectance of the surface should be included with the sky component. In the absence of better information the surface reflectance should be assumed to be 0·2. The sky components on vertical planes parallel to and perpendicular to the window can be found, in a similar manner, by using **13** and **14**. The three sky components are the three component vectors and must be added, as in **9**, to obtain the illumination vector. Internally reflected light is assumed to contribute only to the symmetrical solid and so does not affect the illumination vector.

15 is used to predict the scalar sky component to which the internally reflected component, IRC, must be added. The internally reflected component depends upon the shape of the room (room index), its size and the reflectances of its surfaces as well as the dimensions of the window and any outdoor obstructions, and these parameters are used to find the values of three factors, b, v and e, from Tables IV, V, and VI. The internally reflected component is given by

$$IRC = b.SC_h + v.e.(g/f)$$

where (g/f) is the glass/floor area ratio.

In this way the scalar daylight factor, the vector direction and the vector/scalar ratio due to daylight can be predicted. Where directional lighting is to be achieved by spotlights or other artificial sources a similar approach may be used by calculating on a point-by-point basis.

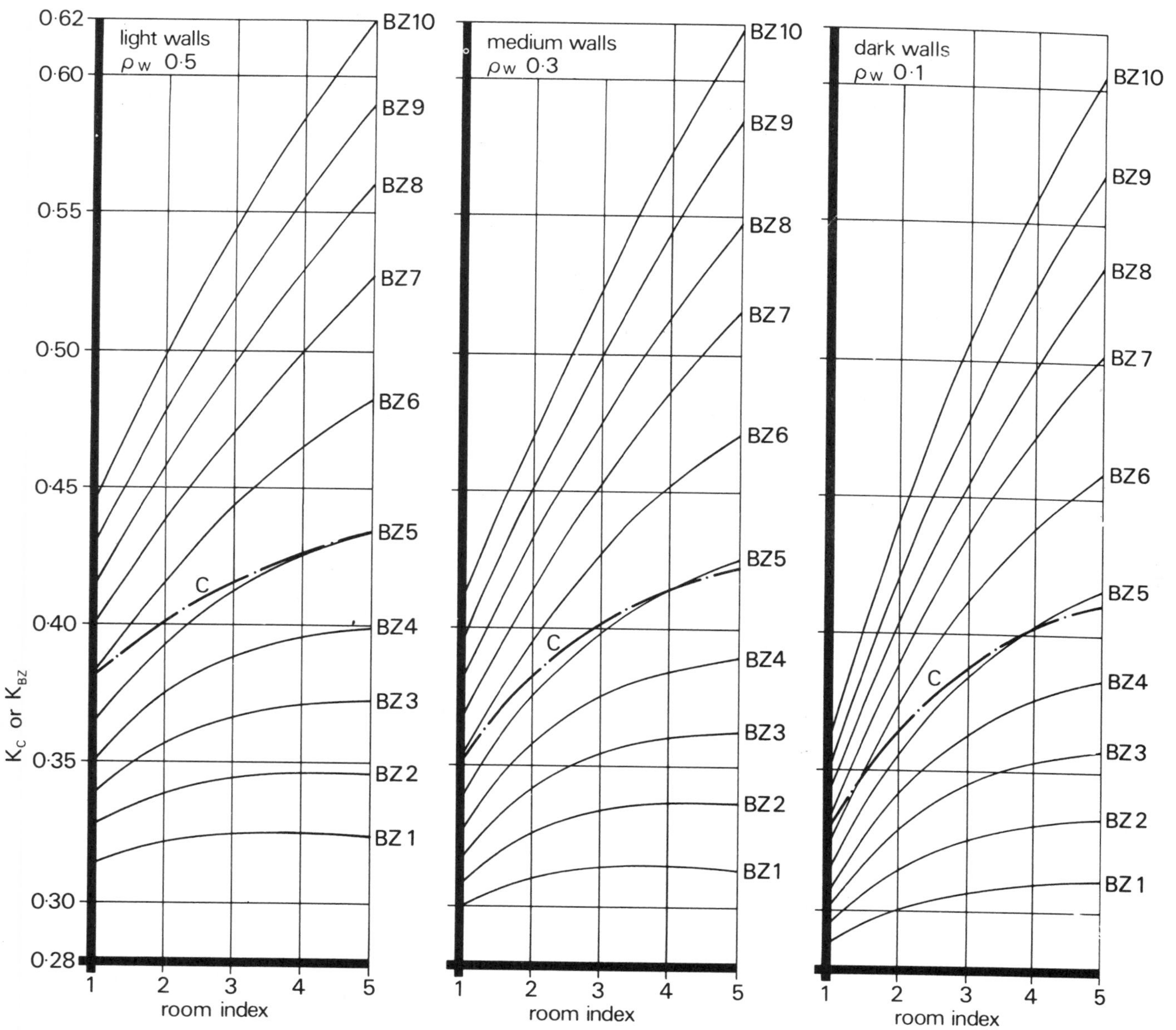

10 *Values of K_{BZ} and K_C for three values of wall reflectance, ρ_w. K_{BZ} is given by the appropriate BZ curve, K_C by the curve C.*

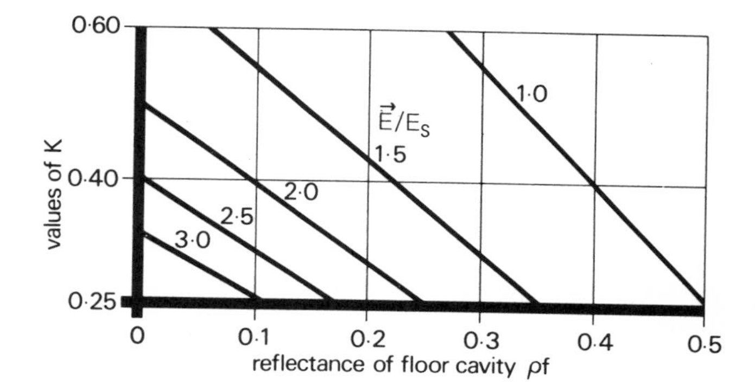

11 *The relation of K to vector/scalar ratio.*

Table III. BZ classification of some typical rooflights. Spacing to height ratios are those recommended as the maximum for uniform lighting.

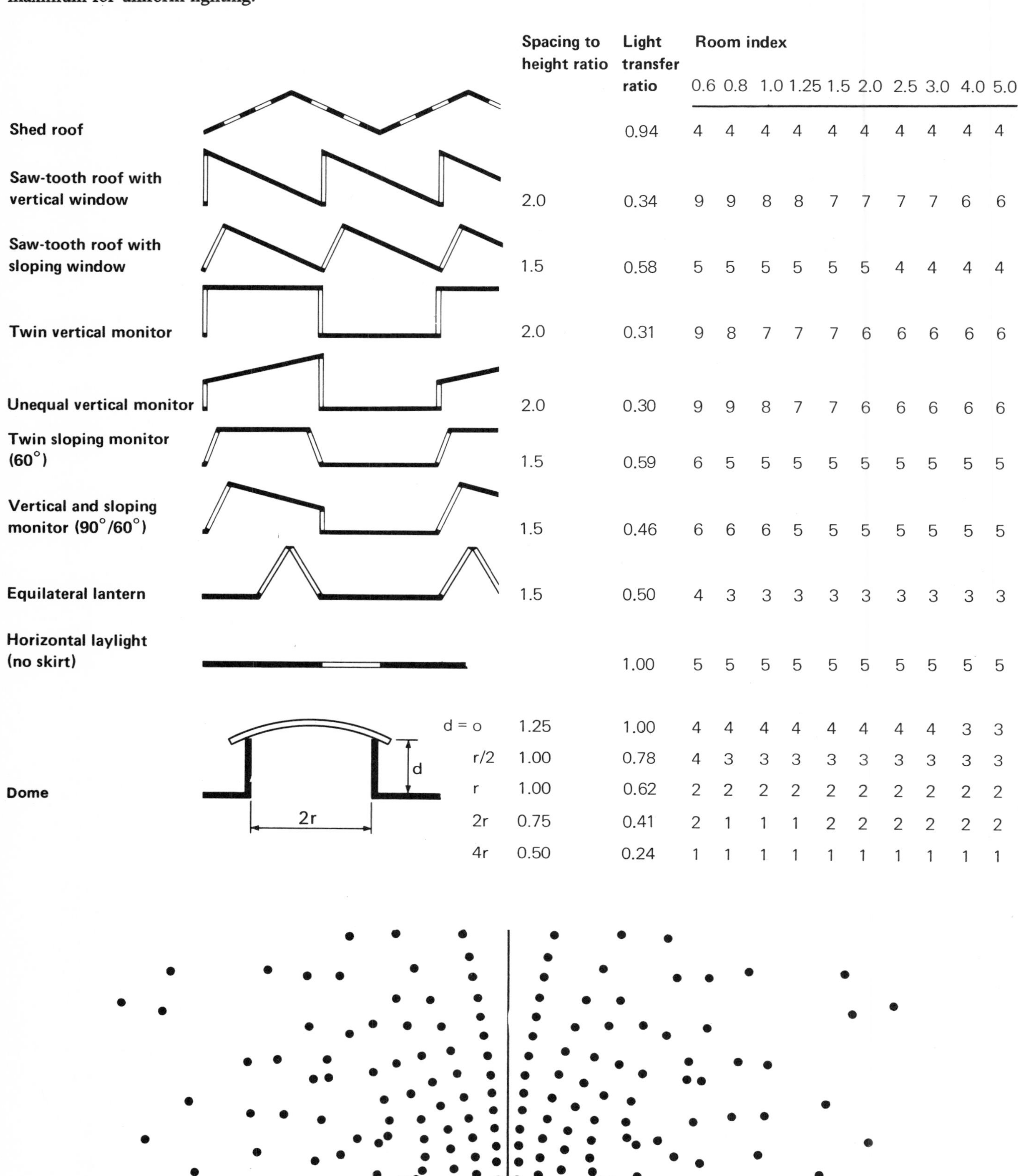

	Spacing to height ratio	Light transfer ratio	Room index 0.6	0.8	1.0	1.25	1.5	2.0	2.5	3.0	4.0	5.0
Shed roof		0.94	4	4	4	4	4	4	4	4	4	4
Saw-tooth roof with vertical window	2.0	0.34	9	9	8	8	7	7	7	7	6	6
Saw-tooth roof with sloping window	1.5	0.58	5	5	5	5	5	5	4	4	4	4
Twin vertical monitor	2.0	0.31	9	8	7	7	7	6	6	6	6	6
Unequal vertical monitor	2.0	0.30	9	9	8	7	7	6	6	6	6	6
Twin sloping monitor (60°)	1.5	0.59	6	5	5	5	5	5	5	5	5	5
Vertical and sloping monitor (90°/60°)	1.5	0.46	6	6	6	5	5	5	5	5	5	5
Equilateral lantern	1.5	0.50	4	3	3	3	3	3	3	3	3	3
Horizontal laylight (no skirt)		1.00	5	5	5	5	5	5	5	5	5	5
Dome d = o	1.25	1.00	4	4	4	4	4	4	4	4	3	3
Dome r/2	1.00	0.78	4	3	3	3	3	3	3	3	3	3
Dome r	1.00	0.62	2	2	2	2	2	2	2	2	2	2
Dome 2r	0.75	0.41	2	1	1	1	2	2	2	2	2	2
Dome 4r	0.50	0.24	1	1	1	1	1	1	1	1	1	1

12 *Diagram for sky component on a horizontal plane for a C.I.E. sky and for vertical glass with a direct visible transmittance of* 0·87

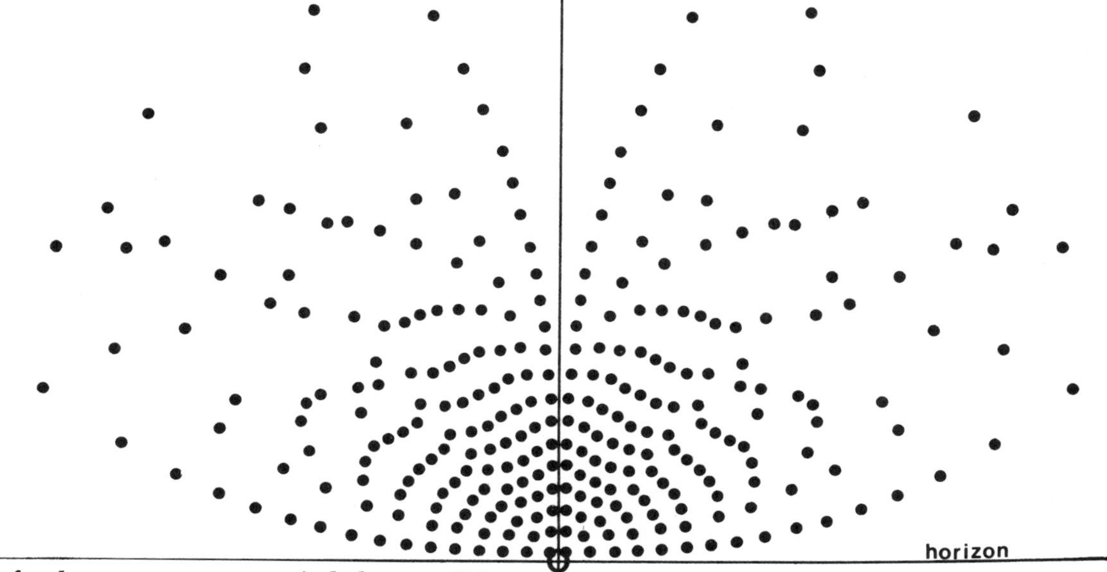

13 *Diagram for sky component on a vertical plane parallel to the window for a C.I.E. sky and for vertical glass with a direct visible transmittance of* 0·87.

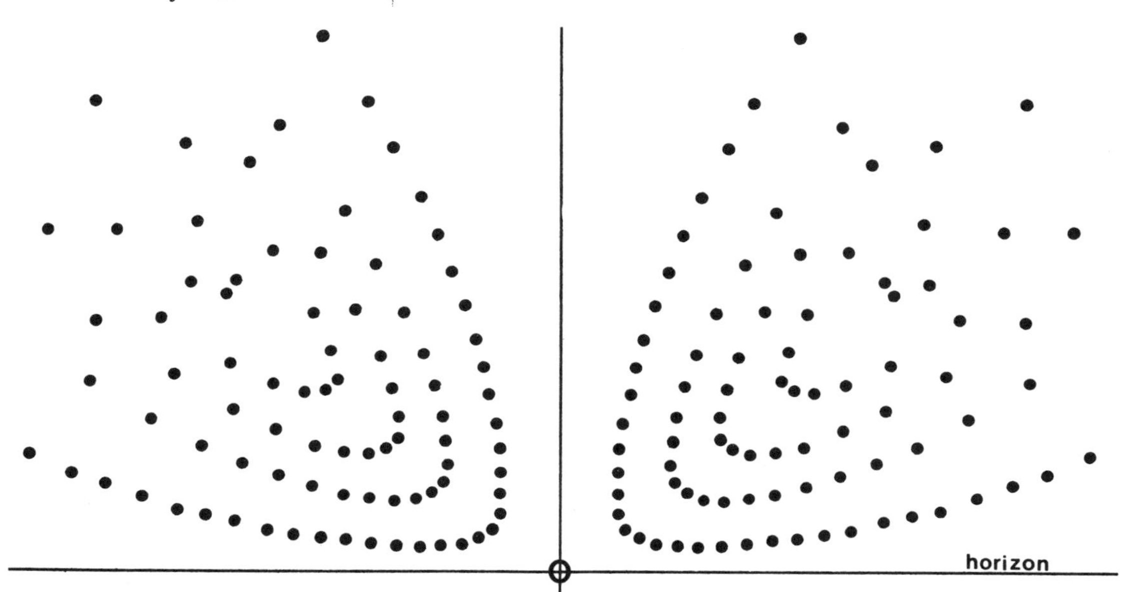

14 *Diagram for sky component on a vertical plane perpendicular to the window for a C.I.E. sky and for vertical glass with a direct visible transmittance of* 0·87.

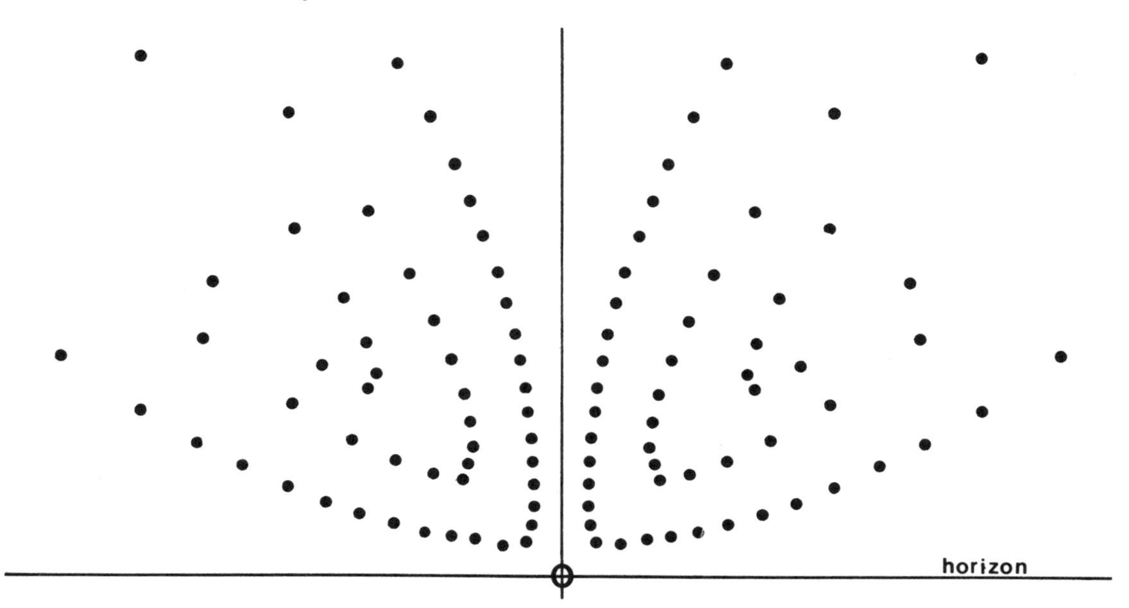

15 *Diagram for scalar sky component for a C.I.E. sky and for vertical glass with a direct visible transmittance of* 0·87.

Table IV. Values of b*

Reflectances											
Floor	0·3			0·1							
Ceiling	0·7			0·7			0·5			0·3	
Walls	0·5	0·3	0·1	0·5	0·3	0·1	0·5	0·3	0·1	0·3	0·1
Room Index	*Values of b*										
1·0	0·1	0	0	0·1	0	0	0	0	0	0	0
1·25	0·2	0·1	0	0·1	0·1	0	0	0	0	0	0
1·5	0·3	0·2	0·1	0·3	0·2	0	0·1	0	0	0	0
2·0	0·4	0·3	0·2	0·4	0·3	0·1	0·1	0·1	0·1	0·1	0
2·5	0·6	0·5	0·3	0·5	0·4	0·2	0·2	0·2	0·1	0·1	0
3·0	0·8	0·6	0·4	0·6	0·5	0·2	0·3	0·3	0·1	0·2	0·1
4·0	1·1	0·9	0·6	0·9	0·7	0·4	0·4	0·4	0·2	0·3	0·2
5·0	1·4	1·2	0·7	1·1	0·9	0·5	0·6	0·6	0·3	0·5	0·3

* Note that this table, which is used in calculating internally reflected components, is not the same as Table I in paragraph 4.10 of Chapter 2, which is used in calculating daylight factors.

Table V. Values of v.

Reflectances											
Floor	0·3			0·1							
Ceiling	0·7			0·7			0·5			0·3	
Walls	0·5	0·3	0·1	0·5	0·3	0·1	0·5	0·3	0·1	0·3	0·1
Room Index	*Values of v (C.I.E. Sky)*										
1·0	3·9	3·2	1·4	3·2	2·4	1·3	2·2	1·8	1·0	1·4	0·8
1·25	3·8	3·1	1·4	3·1	2·3	1·3	2·1	1·7	1·0	1·4	0·7
1·5	3·7	3·0	1·3	3·0	2·2	1·2	2·0	1·7	1·0	1·3	0·6
2·0	3·5	2·8	1·3	2·8	2·0	1·1	1·9	1·6	0·9	1·2	0·4
2·5	3·3	2·6	1·2	2·6	1·8	1·0	1·8	1·4	0·8	1·1	0·4
3·0	3·1	2·4	1·1	2·4	1·7	1·0	1·7	1·3	0·8	1·0	0·3
4·0	2·6	2·0	1·0	2·0	1·3	0·8	1·5	1·1	0·7	0·8	0·3
5·0	2·2	1·5	0·9	1·6	1·0	0·7	1·2	0·9	0·6	0·6	0·3

Table VI. Values of e.

α	e	h/d	e
0	1·00	0	1·00
10	0·88	0·2	0·87
20	0·74	0·4	0·71
30	0·56	0·6	0·55
40	0·39	0·8	0·42
45	0·31	1·0	0·31
50	0·23	1·5	0·14
60	0·10	2·0	0·07

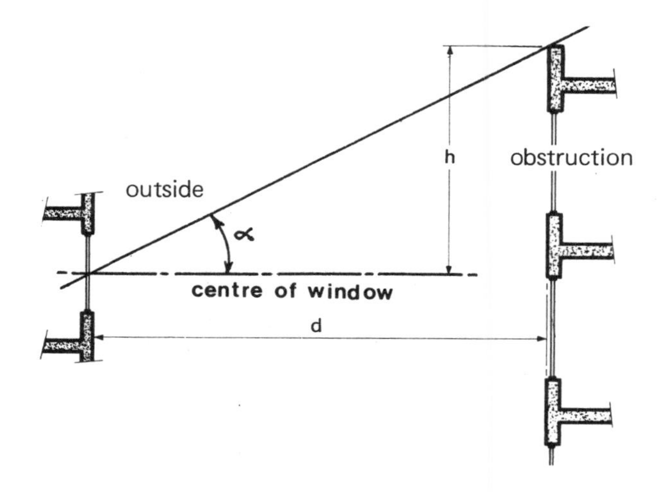

References

1 Lynes, J. A., Burt, W., Jackson, G. K. and Cuttle, C. 'The flow of light into buildings'. *Trans. Illum. Eng. Soc. London.* Vol 31, no 3, 1966. Pages 65-91.

2 The Illuminating Engineering Society. *The IES Code for interior lighting.* London. January 1973.

3 British Standards BS CP3, Chapter 1, Part 1. *Code of basic data for the design of buildings.* 'Daylighting'. London. British Standards Institution. 1950.

4 Cuttle, C., Valentine, W. B., Lynes, J. A. and Burt, W. *Beyond the working plane.* Paper P-67.12 in the Proceedings of the CIE Conference, Washington, 1967.

5 Cuttle, C. 'Lighting patterns and the flow of light'. *Lighting Research and Technology.* Vol 3 no 3, 1971.

6 The Illuminating Engineering Society. *The calculation of utilization factors—the BZ method.* IES Technical Report No 2. London, February 1971.

7 Illuminating Engineering Society. *Daytime lighting in buildings.* IES Technical Report No 4, second edition. London 1972.

8 Hopkinson, R. G. and Longmore, J. 'The permanent supplementary artificial lighting of interiors'. *Trans. Illum. Eng. Soc.* London 1959.

9 *Windows and environment.* Pilkington Brothers Limited, 1969.

Part I Window Design

5 Window design for non-temperate climates

The design of windows for non-temperate climates, in common with the design of other building elements, should be determined by the climatic conditions that prevail at the locality of the building. Observation of local building form will often give useful guidance and whenever local climatic data are available they should be used. The problem facing the designer is that reliable recorded design data for many of the non-temperate areas of the world are frequently not available, although the situation is improving. This chapter classifies climates in identifiable types so that, by relating the type of area for which the design is required to the corresponding climatic type, the appropriate design conditions can be determined.

The types of climate considered are:

1 High latitude temperate *a westerlies*
b continental
2 Middle latitude temperate *a coastal Mediterranean*
b plateau
3 Sub-tropical, tropical arid and semi-arid
4 Monsoon
5 Maritime trade wind
6 Equatorial *a true equatorial*
b with monsoon characteristics

1a High latitude temperate—westerlies
Overcast skies are frequent all the year round. The ground is usually dull in colour and covered with vegetation. The British Isles are typical.

1b High latitude temperate—continental
Overcast skies are frequent in winter but may be clear for lengthy periods. The ground is usually covered in winter snow. In summer the skies are usually clear. The ground in summer is usually dry dusty often bright. High summer temperatures are often experienced. Central France and Germany are typical.
The temperate climate with its overcast skies in the winter months poses problems to the window designer because of difficulties in designing windows to admit sufficient daylight on dull overcast days in winter whilst avoiding summertime overheating. Some sunshine is welcomed all the year round.

2a Middle latitude temperate—coastal Mediterranean
Clear, almost cloudless skies are experienced in summer. There is considerable sunshine in winter though there are some overcast days. There is little rain in summer with the result that the ground is usually bright and highly reflective.

2b Middle latitude temperate—plateau
Exceptionally clear deep blue skies of low brightness are usual. Direct sunshine is intense and reflected illumination and solar radiation often high due to high ground reflectance.
The middle latitude temperate climate with its predominantly clear skies makes it unreliable to base window design for daylight on the overcast sky. Diffuse radiation from the sky, which is often deep blue, is frequently small. Shading of some form is usually necessary to exclude direct sunlight and to moderate glaring reflections from the ground and other sunlit surfaces. These areas are the birthplace of the venetian blind and louvered shutter. Sufficient illumination for interiors may be obtained by utilising the light which is reflected through the slits in the shutters. The architecture in these areas has been strongly influenced by the need to utilise reflected natural light and at the same time exercise control over direct radiation.

3 Sub-tropical, tropical arid and semi-arid
Continuous sunshine occurs practically all the year round. The ground is bright and highly reflective. There is little rainfall and overcast conditions are extremely rare. The Sahara, Egypt, the Middle East and North Africa are typical. Traditionally these areas use small windows in thick heavy wall construction, often of masonry. Dwellings are closed during the day to keep the heat out and small apertures help. Windows are placed high in the walls to avoid a direct view of the ground but are able to make use of reflected light from the ground in order to illuminate the interior. The interior illuminance levels are usually low.

4 Monsoon
Clear skies occur throughout part of the year followed by an extremely cloudy period which gradually disperses. The rainfall is seasonal and the ground brightness varies considerably depending on the time of year. Dusty conditions often occur in the dry season. In upland areas the intensity of direct solar radiation may be high but the diffuse radiation is low when the dust level is low. Nairobi is typical.
Here the design requirements for windows are complex because both overcast and bright clear skies occur. Glare and overheating are severe problems at certain times of year. Window design has to be a compromise between lighting and thermal comfort requirements.

5 Maritime trade wind
Clear skies with scattered cumulus clouds occur all the year round. Overcast skies are extremely rare except during tropical cyclones. The terrain causes local variation but the ground is often covered in vegetation and can be bright.

6a True equatorial
Considerable cloud and rain occur all the year round. The ground is always covered with bright green vegetation. Cloud cover increases during the afternoon.

6b Equatorial with monsoon characteristics
Considerable rain and cloud cover but increasing to a peak during the monsoon period. The ground is covered with

vegetation particularly in the wettest season. There is considerable moisture in the atmosphere giving rise to bright skies even under clear conditions.
Window design in the humid areas typical of the Equatorial climate is generally more concerned with ventilation than with illumination. Wide eaves, carried close to ground level, protect the eyes from sky glare. The ventilation requirements, the protection from glare and provision of adequate natural light on dull overcast days result in compromise design.

Window design

In temperate countries it is usual to separate the roles of daylight and sunlight and, as a result, the daylight factor from an overcast sky has become established as a means of designing windows. In non-temperate countries it is frequently irrelevant to design windows on the basis of an overcast sky or on the basis of daylight factor. In temperate countries there are usually hours of work outside the hours of useful daylight; in non-temperate areas this situation exists less and in the tropics it is rare for working hours to exist outside the periods of available natural light.
In the tropics the sun tends to be higher in the sky and the average outdoor illuminance is higher. The available solar radiant heat is also greater. This all means that window design will be influenced by the need to cope with higher levels of outdoor illuminance, greater probability of glare, and greater amounts of solar radiant heat and therefore increased risk of overheating problems than would normally be the case in more moderate climatic regions.
In dry tropical regions outdoor illuminances of 20 000 lux can be assumed and in hot humid regions 10 000 lux would be likely to occur. (In the UK 5000 lux is considered an appropriate design level.)
Indiscriminate use of glazing is often the cause of high heat gains and losses in buildings. The insulation properties of single glazing are low compared with many other building materials and unless special heat rejecting glasses are used solar radiation is transmitted with very little loss in heat energy. The use of special glasses can make significant reductions in heat loss and gain but their success will depend, in addition to the absolute performance of the materials themselves, on the proportion of the problem that is associated with the glazing.
The heat entering a building through windows during the day not only warms the inside air but also warms the various surfaces of the room. In addition to this effect, the temperature of the glazing increases and causes an increase in the mean radiant temperature within the building. Whilst a gain of heat may be desirable during cold winter months it may be much less desirable during summer conditions. It is obvious therefore that some optimisation of window area is desirable, having due regard for the external climatic conditions that obtain and the internal environmental conditions that must be maintained.
The optimisation can take the form of adjustments to area and shape or type of glazing material employed. A wide range of glasses is available. Table XII, Chapter 1, Function 8, shows the range of performances. In addition to restricting the area of glazing or the use of special glazing materials various forms of shading, preferably external, may be employed. By using external shading the solar radiation is prevented from reaching the window thus reducing the rise in mean radiant temperature associated with any rise in temperature of the glazing itself. To prevent the shading device itself from becoming a heat source it is desirable to use a light coloured surface finish.
Internal shading devices can be successfully employed in situations where the overheating problem is of short duration. The energy absorbed by the device becomes a problem within the interior if prolonged exposure occurs.
The aim in designing windows for buildings that are to be air conditioned should be to achieve a good quality of daylighting without the use of massive areas of external unshaded glazing otherwise the heat gains to the interior will be considerable. This is of particular importance in dry climates where the ground tends to be highly reflective. In such areas small openings will provide an adequate level of illuminance especially if internal finishes are light coloured and the windows are designed with care. In humid climates the openings normally required for ventilation will not usually be needed in an air conditioned building. Provision must be made for natural ventilation because of the risk of breakdown in plant. Opaque shutters can be used to seal any apertures that are additional to daylighting apertures. If light coloured they will assist daylighting by interreflection.
It is advisable to avoid large areas of wall and window that are difficult to shade and are exposed to solar radiation for long periods. Glazing set deep in the structure will often be successful in areas such as the Middle East where the sun is predominantly high in the sky except at sunrise and sunset.
The planning of internal spaces according to use and environmental considerations should be done with great care and done early in the design process. When possible, advantage should be taken of the heating effects of sunlight to achieve economy of operation but, in hot climates, the requirement will more often be to avoid excessive heat gain.
When shading devices are used they should be external. As an alternative, or supplement, to shading the use of special solar control glasses should be considered.
The use of double glazing may well be justified in hot dry climates where the conduction gains can be considerable, especially if large windows are to be used. In hot humid climates the case for double glazing is less for the reason that the interior/exterior temperature differences are usually small and the air is conditioned by dehumidification rather than by achieving a large drop in temperature.

Part II Energy conservation

6 A low-energy approach to office lighting*

The IES has recently given a clear statement of its position regarding lighting and the growing need for energy conservation. This is no policy of surrender: better use of our energy resources is not to be achieved by reverting to the practices of earlier times. Instead, the underlying theme of the statement is that the role of the Society in a changing situation must be to direct our technology and engineering skill to 'facilitate a high level of productivity and a working environment which satisfies current expectations, with minimum use of energy'.[1]

The study† described in this article evolved during a discussion of how the role of daylight in modern offices might be influenced by increased energy consciousness in building design. It was apparent that the only way to develop a worthwhile understanding of the practical implications would be to tackle them by developing an actual installation. The target was to equal, or better still to surpass, by using a low-energy solution, the standards of provision achieved by the best present practice. This was taken to be a permanent overall illuminance of 1000 lux on the horizontal working plane from low glare ceiling mounted luminaires, which typically would operate at an electrical loading in the region of 35-45 W/m^2 of floor area. To be regarded as successful, the low energy solution must match such an installation in both the task visibility it provides and the user satisfaction it achieves, in addition, of course, to using substantially less energy.

Three design objectives were thus defined:

1 to achieve levels of task visibility comparable with advanced current practice;

2 to consume significantly less energy than is general in current practice; and

3 to be judged by office staff to be an attractive alternative to current practice.

Task visibility

When an office worker performs the common task of reading black print on white paper, the purpose of the lighting is to reveal the contrast in the task. Increasing the task luminance raises the reader's contrast sensitivity and so increases task visibility, but this beneficial effect will be partially counteracted if the additional illuminance also produces veiling reflections. Printing ink has a fairly high degree of gloss, as do many types of paper, and so if a light source is at or near the mirror angle the resulting veiling reflections will dilute the observed contrast between paper and ink. The CIE has published a procedure[2] for evaluating lighting according to the task visibility it achieves, and some features of this system are summarised in Table I.

Measurements by Sampson[3] indicate that conventional office lighting tends to have low lighting effectiveness factors (LEF), typically around 0·5. The principal reason for this is that to restrict direct glare the emitted light is concentrated in a near vertical downward direction so that the luminaires have low brightness at normal viewing angles. The effect of this distribution on veiling reflections can easily be visualised by imagining the effect of replacing the office worker's paperwork with a plane mirror: the luminaires would seem to be very bright. Thus a modern low-glare lighting installation providing an average illuminance of 1000 lux on the horizontal working plane may typically achieve an equivalent sphere illuminance (ESI) of only 500 lux.

To achieve a high LEF it is necessary to minimise veiling reflections, either by suitable polarisation of the incident light or by arranging a suitable spatial distribution of the incident light. Polarisation was not pursued, but it became evident that worthwhile gains could be achieved by providing light principally from each side of the subject, so that the ceiling and upper walls as seen from the task position have relatively low luminances.

Calculations showed that two 20 W fluorescent tubes mounted over opposite ends of a standard office desk could achieve a planar illuminance of 260 lux at the centre. Sampson's data suggested that for such an arrangement an LEF in the region of 2·0 could be expected, giving an ESI of 520 lux. Some relatively low background lighting would be necessary to achieve a satisfactory distribution of brightnesses with the space, and this would further increase task visibility.

By separating the lighting in this way to provide carefully controlled task lighting of high lighting effectiveness factor together with a modest level of background lighting (or, to use Hopkinson's[4] term, building lighting), there was a real prospect of achieving the first two design objectives.

Energy consumption

If the density of occupation is one office worker per 10 m^2 of floor space and the lighting electrical loading is 40 W/m^2 (excluding that required by the air conditioning plant to remove the heat generated), the electrical loading directly attributable to lighting is some 400 W per occupant. Alternatively, if each office worker's task lighting is provided by two 20 W fluorescent tubes, allowing for control gear losses the electrical loading is 54 W per occupant. To this must be added some background lighting to ensure safe movement and a suitable gradient of brightnesses. It seemed probable that the installed lighting load could be in the region of 100 W per occupant.

Measurements in offices with different levels of lighting have shown that lighting usage increases when artificial lighting levels are raised, or daylighting is reduced.[5] Whereas the 1000 lux lighting installation is designed for continuous operation during working hours, the relatively low adaptation condition of the local-plus-background lighting would probably mean

* This article was first published in *Light and Lighting* and *Environmental Design*, January/February 1975.

† The study is registered as a contribution towards the long-life/loose-fit/low-energy project initiated by Alex Gordon during his year as President of the RIBA.

1 *General view of office showing three visual task areas—desks, plans and drawing boards:* **above,** *on an overcast day task lighting is switched on but background lighting is provided by daylight;* **below,** *as daylight fails both task lighting and artificial background lighting are switched on. Note variety of brightnesses throughout the room, particularly when background illumination is produced by daylight.*

that in offices with reasonable fenestration it would be the exception rather than the rule for the entire lighting installation to be in use. Task lighting would be under the individual control of the occupants, who would be free to decide whether or not the daylight illumination was sufficient for their work, and the background lighting would be under automatic control to switch on when the daylight fell below some appropriate

2 *View of drawing boards and illustrators' work areas:* **above,** *on an overcast day task lighting with daylight background;* **below,** *as daylight fails artificial task and background lighting. Note parallel motion arrangement beneath board to keep adjustable luminaires mounted vertically when board angle is changed. Also note ink drying lamp being used by illustrator on left above.*

level. In this way fenestration that was not capable of providing adequate task illuminance for more than a small part of the working day could nevertheless be an effective and worthwhile source of background illumination.

The concept of *Task Illuminance Efficacy* (TIE, see Table I) is proposed as a means of comparing the energy performances of lighting installations. Calculations indicate that a system of task

Table I. Terms used in the text

Term	Symbol	Unit	Relationship	Notes
Task illuminance*	E	lux		For general office lighting, taken to be equal to average illuminance of horizontal working plane.
Sphere illuminance†		lux		Illuminance under a standard lighting condition: totally diffuse and unpolarised lighting as received at the centre of an integrating sphere.
Equivalent sphere illuminance	ESI	lux		For an illuminated task, the sphere illuminance that would achieve the same measured visibility (although the unit is lux, ESI specifies the visibility of a task, not its illuminance).
Lighting effectiveness factor†	LEF		$\text{LEF} = \frac{\text{ESI}}{\text{E}}$	An index of the effectiveness of illuminance to achieve task visibility.
Overall luminous efficacy	(lm/W)	lm/W		Ratio of lamp service lumens to lamp circuit watts.
Lighting electrical loading	(W/m²)	W/m²		Lighting electrical load per square metre of floor area.
Utilisation factor*	UF		$\text{UF} = \frac{\text{E}}{(\text{lm/W}) \times (\text{W/m}^2)}$	Proportion of lamp lumens incident on horizontal working plane.
Task illuminance efficacy	TIE	lm/W	$\text{TIE} = (\text{lm/W}) \times \text{LEF} \times \text{UF} = \frac{\text{ESI}}{(\text{W/m}^2)}$	An index of the energy performance of a lighting installation, that relates task visibility achieved to power consumed.

* See *IES Code for interior lighting* (1973). † See *American IES Lighting Handbook* (5th edn.) (1972).

3 *View of desk area:* **above,** *on an overcast day, task lighting with daylight background;* **below,** *as daylight fails artificial task and background lighting. Note difference between dark unoccupied desk in centre and bright occupied desks on either side.*

4 *Close-up view of a single desk;* **above,** *on an overcast day with task lighting only;* **below,** *as daylight fails with artificial task and background lighting. Note how desk surface is brighter than surrounding areas drawing attention to visual task.*

and background lighting might achieve TIE values of more than 50 lm/W as opposed to the 10-20 lm/W achieved by a conventional lighting system: 'High TIE' is a worth while design objective.

Preference

Ten investigations of illuminance preference reviewed by Fischer[6] are used to support the claim that the illuminance that people prefer is in the region of 2000 lux,[7] but perhaps this should more properly be described as the level at which people prefer the conventional type of office lighting. There are other, particularly non-working situations, in which not only are much lower illuminances preferred, but also very different types of lighting installation are usually chosen.

The open question, then, was whether a system of individual task lighting that achieves high standards of visibility and comfort with a modest and not necessarily uniform level of background lighting would be preferred to a conventional installation that provides higher illuminance. The way to find out was to install this type of lighting, so that the reactions of office staff could be obtained.

Trial installation

The office of the Pilkington Environmental Advisory Service is an open area of 210 m^2 with a floor to ceiling height of 3·1 m. Windows in three walls give an average daylight factor greater than 3 per cent, but falling to ½ per cent in some places. An average illuminance of 600 lux was provided by recessed fluorescent luminaires with flush prismatic panels. **1** gives a general impression of the area.

The nine members of the advisory staff each have a desk, but there are three plan tables in the centre of the space for dealing with large building drawings. In addition, there are three technical illustrators working at drawing boards, and two secretarial staff fulfilling the usual range of typing, filing and receptionist duties. The work is probably more varied and more visually demanding than average.

The task lighting for the desks was developed in a prototype situation. Various reflector types and mounting positions were studied, and several members of the EAS staff co-operated at this stage. No suitable commercial fitting was found and so for the trial installation a special reflector was made, incorporating a shield beneath the tube to reduce specular reflections on the desk top at each side of the central task area. The height of the reflectors is fixed, but the supports permit some degree of tilt to allow for differences of subjects' eye levels and differences of task. This was important for the secretarial staff where the typewriter places the visual task well above the desk surface.

To evaluate the effectiveness of this lighting, a visibility meter[8] was constructed, and calibrated under sphere illumination using a standard printed paper task. For measurement purposes, a desk-top reference point was specified, directly in front of the seating position (which is offset to the left by the drawer unit) and viewed at 30° from the vertical, this direction having been found to be representative of office tasks.[9]

For the original installation, the average reference point illuminance for the nine desks was 550 lux, and the average ESI was 330 lux, giving an overall lighting effectiveness factor of 0·6. On its own, the new task lighting achieved 280 lux planar and 570 lux ESI (LEF = 2·04), the ESI decreasing to around 450 lux as the task was displaced towards the edges of the desk. The contribution of the background lighting was rather greater than expected. Almost three-quarters of the tubes of the original installation had been removed but its effect was to raise the average reference point illuminance to 410 lux, and the average ESI to 740 lux, giving an overall lighting effectiveness factor of 1·80. **1b** and **3b** show that the background lighting is concentrated around the office perimeter so that it provides both preferential illuminance in main task areas and relatively bright walls, these emerging as the principal factors affecting assessments of the adequacy of the background lighting. The desks face towards the less illuminated middle of the office, so that the lighting is principally from above or slightly behind the desk worker, and so less likely to produce veiling reflections than a uniform arrangement. Neither the furniture nor the luminaires were relocated for the study.

The plan tables required a different lighting solution as obstructions caused by the reflector supports would not be acceptable. Overhead lighting from a fluorescent ceiling mounted luminaire with a metalised plastics louvre employing cells of parabolic cross-section provides 600 lux on the plan table surface. However, as the lighting is concentrated downwards as with conventional low-glare overhead lighting, veiling reflections are present, and appear more evident by comparison with the desk situations where they are virtually eliminated. The use of transparent overlays on drawings has been noted as one type of visual task that is appreciably easier to perform on the desks than on the plan tables. The plan table lighting is controlled by cord-pull switches.

Each drawing board is illuminated by two adjustable fluorescent units each housing two 15 W fluorescent tubes. A special parallel motion arrangement was designed and installed so that as the board angle is changed the mounting pivot of each unit remains vertical so that the spring loaded arms are not set out of balance. Each of the illustrators appears to have evolved his own idea of the arrangement of the lighting units that suits him best and it remains to be seen whether in due course a concensus pattern will emerge. Spot checks with a light meter indicate that the units are set to achieve illuminances not very different from the 600 lux provided by the previous lighting installation, but despite this two of the illustrators had previously imported 'Anglepoise' lamps to supplement the lighting. The banishment of these low efficacy sources prompted the predictable complaint from the illustrators that their ink driers were being taken away. To compensate, a radiant ink drier was developed, **2a**, that houses a 50 W 12 V spotlamp (unfortunately no infra-red source of suitable size is available) and this has proved very popular.

The background lighting has been achieved by the simple expedient of removing many of the tubes from the existing installation. Careful visual assessments were made during evening as daylight faded to achieve a balance that appeared to be about right and for which the background lighting switched on to supplement the failing daylight at about the right time. For the trial installation a deliberately non-uniform pattern of background lighting has been provided that gives up to 110 lux scalar in the desk areas, but declines to 40 lux scalar at the centre of the office. A photo-electric control is arranged to switch on the background lighting when the daylight illumination in the office falls below a preset level. The photo-electric control is itself controlled by a time clock to prevent it from switching on the lighting out of working hours.

Discussion

Performance data are summarised in Table III, and certainly the desk-top ESI of 740 lux compares well with advanced present practice. The more than fivefold increase in task illuminance efficacy for the desk lighting is notable.

The trial installation was completed in April 1974, shortly before the introduction of British Summer Time, and so the opportunities for the staff to experience the new lighting have been fairly limited. However, it is already evident that the pattern of use of electric lighting has changed substantially. Previously it was the normal procedure for the first person in each morning to switch on the lighting and for the last person out to switch it off. Now that the general lighting is controlled automatically, each individual can decide whether or not he wants to have his own task lighting to supplement the daylight illumination. The availability of daylight in this office is probably above average, but not unusually high, and it is evident that for much of the time people are content to work without electric lighting.

This tends to confirm expectations that the adoption of low energy lighting solutions could result in a reappraisal of the role of daylight in offices. Modern lighting practice has created offices with relatively high standards of task visibility, which should not be sacrificed. In many deep offices daylight penetration is insufficient to meet these standards and so the tendency has been to disregard the contribution which can be made by daylight and to install a high level of artificial lighting. With a low energy lighting system such as that described here, although daylight produces task illumination for part of the working day it can give background illumination for very much longer. It has been found that general background lighting is not needed until daylight illumination is quite low, and so for much of the year this will be experienced only on dull days. Even in winter the daylight illumination should be sufficient for background lighting for part of the working day, but of

Table II. Trial installation lighting loads

	Work area	Type of lighting	Quantity	Power each (W)	Power total (W)
Task lighting	Desks	2 standard luminaires	9	54	486
		1 standard luminaire + 1 adjustable luminaire	2	70	140
	Drawing boards	2 adjustable luminaires	3	80	240
	Illustrators' work areas	1 standard luminaire	2	30	60
		1 adjustable luminaire	1	40	40
	Plan tables	1 ceiling mounted luminaire	3	190	570
Background		Recessed prismatic diffusing luminaires	14	95	1330
Total lighting load					2866

Table III. Comparison of original and trial installations

Installation	Original	Trial
Total lighting load (W)	5130	2866
Lighting load per unit floor area (W/m²)	24·4	13·6
Density of occupation of office (m²/person)	15	15
Background lighting load per occupant (W)	—	95
Total lighting load per desk worker (W)	365	95+54=149
Total lighting load per illustrator (W)	365	95+80+30=205
Average illuminance at working position on desk top (lux)	550	410
Average ESI at working position on desk top (lux)	330	740
Lighting effectiveness factor for desk tasks	0·60	1·80
Task illuminance efficacy for desk tasks (lm/W)	13·5	74·5

course the reduced use of electricity for lighting must be assessed in relation to the various thermal effects of the windows, which will vary from one situation to another.

In the trial installation the advisory staff spend a fair proportion of their working time outside the office so that a third or more of the desks may be unoccupied at any one time. **3** shows how the appearance of the whole office responds to the extent of occupation and the level of activity, as each member of the staff marks his presence with his personal pool of light. If all the lights in the office are switched on, the lighting load is about half that before the low energy system was installed. However, because the lights are only switched on when they are needed the lighting load varies, in contrast with conventional high energy lighting where the load is constant. Thus in an air-conditioned building the peak cooling load is further reduced because the lighting load is a minimum when daylight is high, the time of greatest external heat gains to the building; when the daylight is dull, a time of low external heat gains, the lights are switched on, contributing somewhat towards the heating of the building. Thus the adoption of a low energy lighting scheme can influence the whole energy use of the building.

All these factors point to a substantial potential energy saving. The trial installation is fully metered, and energy use will be recorded for at least a full year.

Questionnaires completed so far by the office workers suggest a generally favourable reaction to the new system of lighting. The subjects are not typical office workers, but whereas one can on the one hand hypothesise that their well above average educational attainment and concern for environmental problems could make them more favourably disposed towards a study of this sort, on the other hand the fact that their day-to-day work brings them into contact with the latest developments in environmental engineering could make them more than usually critical. However, it must be admitted that the original installation is barely representative of advanced practice.

The careful planning of lighting where each working position is catered for separately involves far more attention to detail at the design stage than does the blanket approach of high overall lighting. Background illumination may be supplied as a general uniform level and incorporated into the building structure in much the same way as present practice. Organisational flexibility could then be achieved by each work station having its own built-in lighting so that the limitation on flexibility is the availability of electrical supplies. In fact the increased use of electrical and electronic devices in offices is already demanding a flexible system of mains electricity supply. Thus, although the background lighting is part of the fabric of the building, the task lighting may come to be regarded as part of the furniture, which under present UK tax legislation gives some advantage to the office user.

The provision of the background lighting by selective removal of tubes from the existing installation is obviously less than satisfactory. For a new installation it would be advantageous to eliminate all of the lighting services from the ceiling, and experiments have been carried out to provide indirect background lighting by placing inverted 'downlighting' units using 80 W MBF/U lamps on filing cabinets and other furniture. The pools of light produced on the ceiling could provide another interesting aspect of variety, but their colour during the run-up period was criticised. These experiments will continue.

The ultimate purpose of this study is not to propose that a particular arrangement of electric light sources is the right way to light offices, but rather to demonstrate that a viable alternative to the current trend of lighting practice is feasible, and that reduced energy consumption for office lighting does not mean reduced standards of task visibility or people being committed to work in dingy and unattractive environments.

References

1 *Light and Lighting*, vol 67, p251, 1974.
2 CIE Publication 19. 1972.
3 Sampson, F. K. Technical Report 4. New York, Educational Facilities Laboratories. 1970.
4 Hopkinson, R. G. *Transactions of the Illuminating Engineering Society*, vol 30, p63, 1965.
5 Milbank, N. O., Dowdall, J. P. and Slater, A. J. *IHVE*, vol 39, p145, 1971.
6 Fischer, D. *Lighting Research and Technology*, vol 2, p150, 1970.
7 Illuminating Engineering Society. *The IES Code for interior lighting*. London, 1973.
8 Slater, A. I. *Lighting Research and Technology*, vol 7, no 1, 1975.
9 Allphin, W. *Illuminating Engineering*, vol 58, p244, 1963.

Part II Energy conservation

7 The influence of glazing on energy consumption in offices

Glazing, because of the complex ways in which it influences energy consumption, demands particularly close attention in the present climate in which the need for energy conservation has become established as a major design parameter. In energy terms glazing has traditionally been judged on the basis of its thermal insulation properties (U-value) alone. Its total performance is however more complex than this approach implies.

It can often be self-defeating to optimise one performance parameter of a building or building component in isolation because the consequences of the optimisation may well be to increase energy consumption in another area. Particularly with glazing it is important not to assess it only in terms of heat losses but to quantify all the aspects in which windows affect energy use. This paper gives the results of some preliminary computer studies of energy consumption in a typical centrally heated, naturally ventilated office. In these studies the full effects of window design have been modelled in an attempt to show the influence which variations in window area may have on total consumption.

Useful solar radiation

As well as losing heat by conduction glazing admits heat to buildings because it is transparent to solar radiation. Designers have perhaps instinctively felt that the amount of solar energy available during the heating season was small enough to be ignored but recent meteorological measurements have shown that this is not so. In fact, subsequent research[1] has shown that, throughout the heating season (September to May), glazing facing south of an east-west axis can gain more energy than it loses and may therefore be a net provider of heat to buildings. However, it does not follow that in practice all the solar heat gained through windows can usefully supplement heating requirements. Particularly at the fringes of the heating season there could be occasions when the heating load is met by internal gains and a small proportion of the solar gains, the remainder being 'wasted' in energy terms.

The amount of solar gains which are wasted depends on a range of factors; for instance time of year, geographical location, orientation, dimensions and thermal properties of the building, glazing area, patterns of occupancy and other heat gains. Another very important factor in determining how useful the solar gains will be is the nature of the heating control system. It is vital that a heating system and its controls can respond quickly to fluctuations in solar input so that excess heat is not produced unnecessarily by the heating system when solar heat is available.

Because of the many variables involved in determining the influence of solar heat gains on heating requirements the situation is best modelled by computer. A program has been developed for this purpose.[2] The program uses solar data and temperature data to calculate the annual heating energy consumption of a building of any given dimensions, thermal properties, internal gains, orientation and location. The only solar heat gains which are considered to supplement heating requirements are those which would not lead to the inside temperature rising above the thermostat temperature. The rest of the solar gains are rejected within the program and it thereby distinguishes between the solar gains which are useful and those which are not. The program computes the heating load for each hour of a day, using average meteorological data. The heat load for a day is computed by integrating the hourly values and the heat load for the whole heating season is computed by integrating the daily values.

Heating energy requirement

For the purposes of this study a 5 metre × 5 metre × 3 metre high office module was selected for analysis. Meteorological data for Kew were used. An inside design temperature of 20°C was selected and typical levels of internal heat gain and typical periods of occupancy for office use were assumed. The ventilation rate was taken to be two fresh air changes per hour. There was one external wall in the module, the opaque part having a U-value of 1·0 W/m^{2}K.

1 shows the way in which the total heating energy consumed varies with the percentage of glazing in the external wall. The units of energy consumption are given in gigajoules per square metre of floor area and assume a gas or oil heating system having an overall thermal efficiency of 66 per cent.

1, which is for single clear glazing, indicates the influence which orientation has on energy use, a south facing module having the lowest consumption. The upper line in **1**, marked 'no sun', shows what the energy consumption would be if one did not consider solar radiation in the calculations. In a sense the 'no sun' line represents the traditional calculation of energy consumption. It is obvious that the traditional approach does not adequately represent the thermal performance of windows.

2 shows the variation in energy consumption with window area for clear double glazing. In this case the slopes of the lines are shallower because the heat loss per unit window area is less than with single glazing.

Lighting

The daylight admitted by windows has potential for saving energy if it is used to supplement the artificial lighting requirement. This is another aspect of window performance which can have a significant effect on a building's total energy use.

It has become increasingly common in recent years to rely on high levels of permanent artificial lighting in offices. Often this has been associated with the trend towards deeper offices which have required high lighting levels to compensate for the lack of daylight. However, in office buildings the artificial lighting has now been identified as the biggest single energy consuming component[3] and it would therefore seem sensible to examine the potential influence that daylight can have in this area.

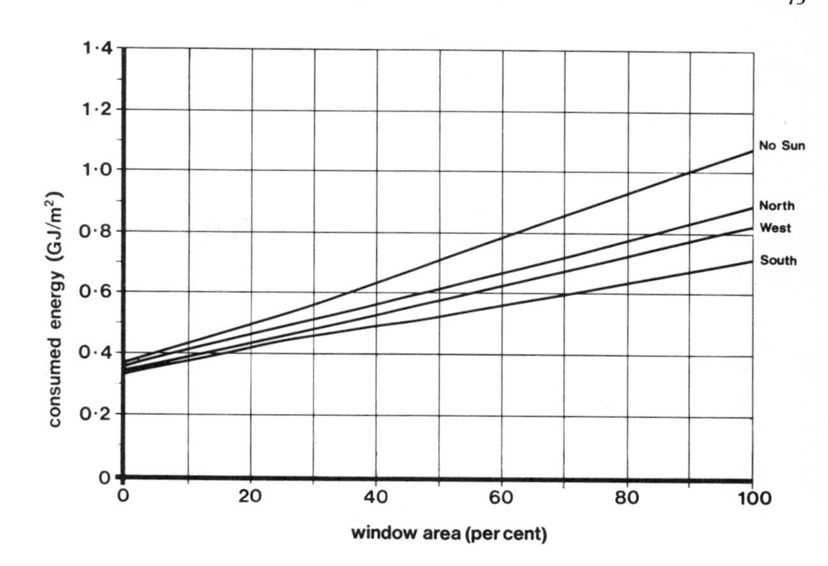

1 *Annual energy due to heating, single glazed office, Kew.*

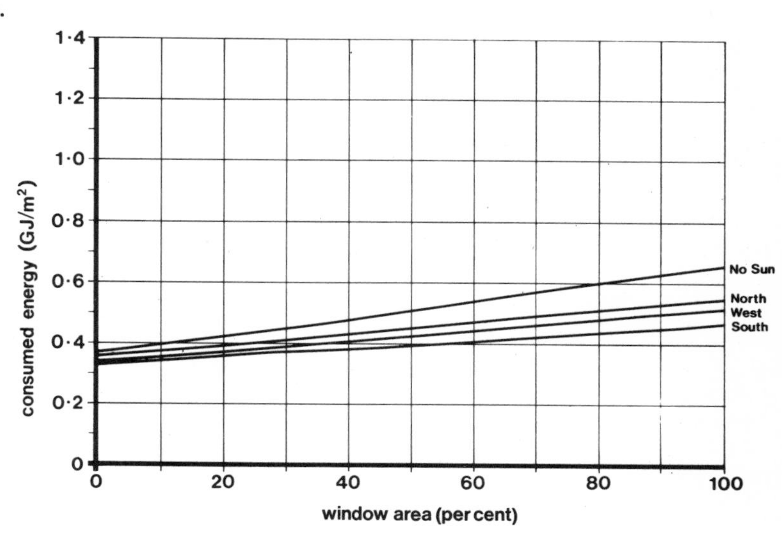

2 *Annual energy due to heating, double glazed office, Kew.*

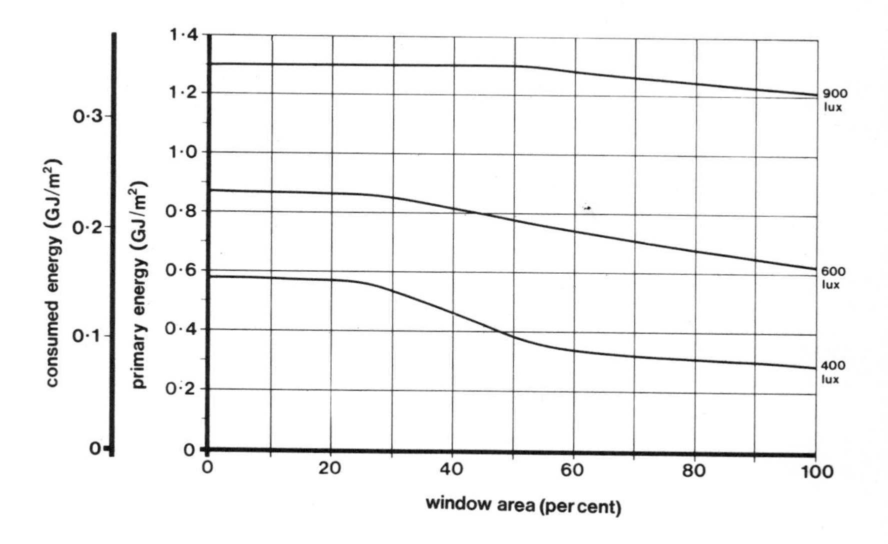

3 *Annual lighting energy.*

The 5 m × 5 m office module has again been used to look at the possible savings in artificial lighting energy as the availability of daylight increases. The IES Technical Report *Daytime lighting in buildings*[4] gives a method of calculating the number of hours per year when daylight reaches various specified levels and hence the number of hours per year when artificial lighting will not be necessary. This method has been used to determine the number of hours per year when the artificial lighting can be switched off in the office module for three different inside illuminance levels.

From the annual hours of use calculated by the IES method the curves shown in **3** have been obtained. Curves are given for three different design levels of uniform overhead lighting. These indicate that the design illuminance is important in determining how much energy can be saved with daylight. For instance, if the internal design level is 900 lux there is unlikely

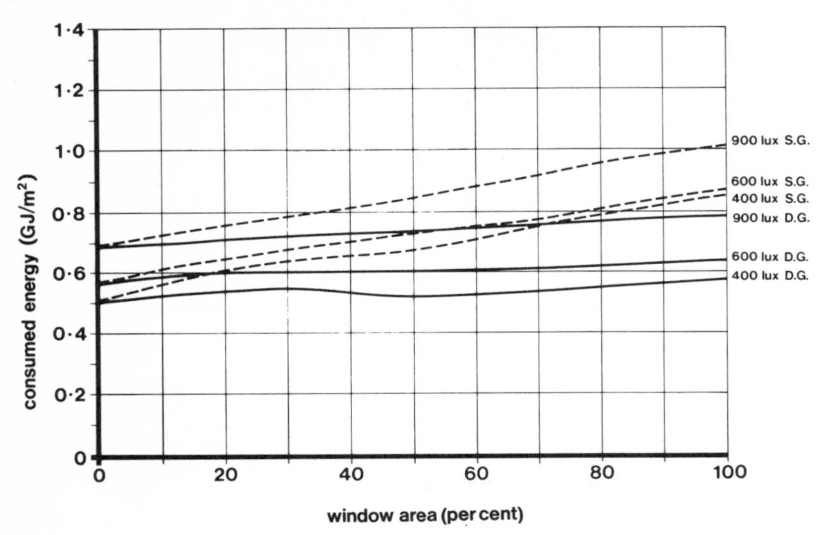

4 Annual energy due to heating and lighting, south facing office, Kew.

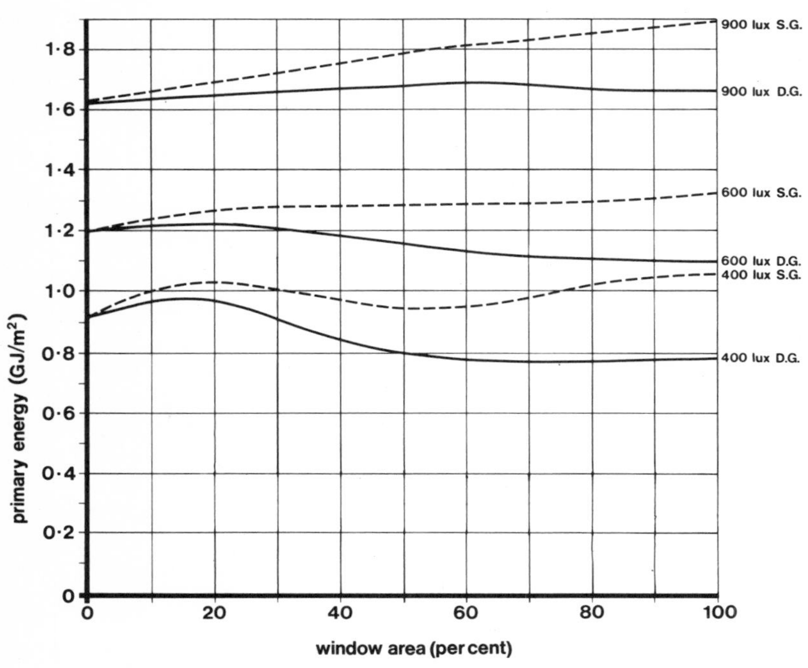

5 Annual energy due to heating and lighting, south facing office, Kew.

to be any artificial lighting energy saved if the window area is below 50 per cent. This is because daylight alone will be unable to provide 900 lux at the back of the office and therefore the electric lights will need to be permanently on. However, if the design illuminance is 400 lux, there will be a much larger proportion of the year when this can be provided by daylight and therefore the number of hours per year when the lights are off will be large.

Primary energy

In **3** the energy consumption is plotted in both 'consumed' and 'primary' energy units. The consumed energy is that which actually reaches the building whereas the primary energy includes the energy overheads required to produce and distribute the fuel. In other words the primary energy content of a fuel reflects its drain on natural resources by including the energy which is lost in generating and distributing it.

The inefficiency in generating and distributing electricity means that 3·73 units of primary energy are required to produce one unit of consumed energy at the building.[5] Hence in **3** the primary energy, which represents the basic resource content of electricity, is 3·73 times the consumed energy. (By way of comparison the primary/consumed energy ratio for oil is 1·08 and for natural gas is 1·06. These fuels therefore lose very little of their energy in distribution and the consumed energy values shown in **1** and **2** will be almost the same as their primary energy requirement.)

The net effect of glazing

The results of the computer program for obtaining heating requirements can be combined with the lighting energy values of **3** to obtain total energy consumption. The net effects are shown in **4** and **5**. In these the benefit of the heat gain from the higher artificial lighting levels has been taken into account when computing the annual heating requirement.

The graphs show how energy consumption for heating and lighting combined can vary with window area. Results are given for office modules with different installed artificial lighting levels and for single glazing (SG) and double glazing (DG). A south facing office is assumed (which is the most favourable

6 Salford Town Hall Extension. Architects: Cruickshank and Seward, Manchester.

orientation), but the total energy consumption will be relatively unaffected by quite large deviations in orientation from due south.

4 and **5** illustrate the difference between consumed energy and primary energy when analysing performance. It has been common until recently to calculate the energy demand of a building in terms of consumed energy. However, because energy costs are now increasingly reflecting the resource costs of a fuel, a more accurate indication of annual fuel costs is probably gained by looking at the primary energy consumption of a building. **5** is probably the more relevant graph to use in terms of running costs and of the demand on natural resources.

Certain assumptions are inherent in producing these results. One is that the heating control system will respond efficiently so as to make use of the solar heat gain through the window. Another is that the artificial lighting will be switched off when daylighting can provide the required illuminance. Both assumptions however are reasonable, given sensible building design and use. Even before the energy crisis it was observed[3] that shallow offices which had reasonable daylight provision would use the artificial lighting about 500 hours per year less than deep plan offices. One can assume that in the present energy-conscious climate the occupants of offices are even more likely to switch off lights when they are not necessary. If a designer remains unconvinced about the value of leaving the control in the hands of the occupants a simple photocell-linked control system for the lighting is a practical alternative. Such a system means that the lights are switched off automatically when a photocell senses a certain daylight level. The relationships shown in **3** should then easily be attainable.

The range of glazing examined in the graphs (namely 0 to 100 per cent) does not of course represent the practical range of fenestration. Very high glazed areas, without shading, are not practical because of summer overheating problems whilst very low glazed areas are not practical because the windows are not then able to provide the essential psychological link with the outside world. However, glazing from 0 to 100 per cent has been covered in this exercise for the sake of academic completeness.

Discussion

The observations to make from a study of this nature should be general rather than specific. When one examines **4** and **5** one sees a series of graphs which on the whole are flat. In other words the net energy consumption of the office module is more or less independent of window area. It appears that the increased heat losses incurred by large windows tend to be counter-balanced by the useful solar heat gains and the savings in artificial lighting energy. As a result the designer is free to vary his window area anywhere between the limits dictated at the lower end by the psychological requirements and at the upper end by the need to maintain acceptable conditions in summer.

5 also gives us some clues about the way offices could be designed to achieve low energy consumption. The curves shown in this theoretical study basically agree with what has been measured in practice; namely that the dominant factor in determining energy use is the installed artificial lighting level. The thermal properties of the building envelope are of secondary importance. The foregoing results have assumed uniform overhead lighting which will be either completely switched on or completely switched off depending on the daylight level at the back of the office. It is, however, perfectly possible to reduce the energy demand of lighting even further by use of some commonsense principles.[6] The first of these is to distinguish between the areas requiring high lighting levels (the desks) and those areas which do not (the space between the desks). Localised lighting appropriate to task illumination can then be provided for the desks and lower levels of background illumination can be provided for the areas between desks. The occupant of the desk could have control over his task lighting so that it is only switched on when he is at his desk and when daylighting is inadequate. Similarly the background lighting can be linked to a photocell-controlled switch so that it is only on when daylight is inadequate for background illumination. The second commonsense principle is therefore to make maximum use of the available daylight. Using these principles an open plan office, which originally had a uniform overhead lighting installation giving 550 lux, has reduced its annual lighting energy consumption by 83 per cent, without detriment to the visual environment.[7]

If daylight, in conjunction with sensible artificial lighting design, has the potential for such large savings it would seem logical to design offices to receive a reasonable contribution of natural light. This immediately imposes a limitation on plan form because an office can be considered to be daylit only up to about 6 m back from a glazed wall. An open office, glazed on opposite sides, can therefore be regarded as daylit for depths up to 12 m. Any increase in depth will inevitably increase the need for electric lighting.

However, a shallow office building has other advantages besides its ability to utilise daylight. Shallow buildings on the whole do not need air-conditioning. Natural or mechanical ventilation is usually adequate for providing fresh air and for keeping temperatures acceptably cool in summer. This is a significant point because measurements have shown that air-conditioned offices use about 50 per cent more energy than non air-conditioned offices.[3]

It appears then that an office building which is long and shallow on plan and has reasonable daylight provision offers the best potential for a low energy building. This indeed is the built form used in the new Salford Town Hall extension, **6**, which has been designed with low energy consumption as a main parameter. The principles of individually controlled task lighting have been used in the open offices so that full benefit is obtained from daylight. The building and its design philosophy are well documented in a paper given at the recent CIB Energy Symposium.[8]

Intuitively one might regard a building of this shape as being wasteful in energy. It certainly conflicts with the basic physics teaching that compact shapes have lowest heat loss per volume of enclosed space. However, heat loss is not the only factor affecting energy consumption. In offices it is not even the major one. It is essential to look at all the energy consuming aspects of building, and the way that they inter-relate, if low energy design is to be achieved. This is particularly so with glazing because of the complex ways in which it influences the energy balance of buildings.

References

1 Wilberforce, R. R. 'The energy balance of glazing'. *Building Services Engineer*, March 1976, pp241-243.
2 Newton, A. 'A practical computer program for predicting energy consumption and other environmental parameters in buildings'. (To be published).
3 Milbank, N. O., Dowdall, J. P. and Slater, A. *Investigation of maintenance and energy costs for services in office buildings*. Building Research Station CP38, 1971.
4 Illuminating Engineering Society. *Daytime lighting in buildings*. IES Technical Report 4. London 1972.
5 Building Research Station Working Party. *Energy conservation: a study of energy consumption in buildings and possible means of saving energy in housing*. Building Research Station CP56, 1975.
6 Cuttle, C. and Slater, A. I. 'A low energy approach to office lighting'. *Light and Lighting*, January/February 1975.
7 Owens, P. G. T. *Energy conservation and office lighting*. CIB Symposium on energy conservation in the built environment. Watford, 1976.
8 Winch, G. R. and Burt, W. *Energy conservation measures in buildings—environmental design strategy*. CIB Symposium on energy conservation in the built environment. Watford, 1976.

Part III Solar control

8 Solar-control performance of blinds

The solar heat gain through windows is determined by the orientation, size, and external shading of the windows and by the properties of the glass and any associated blinds; the effects of the admitted heat are modified by the thermal properties of the building and by the use of ventilating or air conditioning plant. All these interact and contribute to the whole window design as discussed in Chapter 1 but it is convenient to analyse the factors separately and this chapter is concerned with the effects of blinds.

The decision to install blinds and the type of blind selected (horizontal or vertical louvers, roller blinds or curtains) may depend on considerations other than solar control—to provide privacy, to prevent glare or for appearance. Whatever the reason for their use, blinds affect the solar control performance of the windows and an appropriate allowance must be made.

The common criterion of window performance is the shading coefficient[1] (defined in the glossary in Appendix 2 and discussed in Chapter 1, Function 7, paragraph **7.09**) which, for windows with blinds, depends upon the solar optical properties of the glass and the material of the blind, on the coefficients of heat transfer at the window surfaces, and on the geometry of the blind and the angular position of the sun.

The relevant solar optical properties are the reflectance, absorptance and transmittance of the material to radiation of spectral distribution defined by Parry Moon's curve for air mass 2.[2] Some typical values are tabulated in Table I.

Table I. Solar optical properties of typical blind materials and glasses

Blind material	Reflectance	Absorptance	Transmittance
Opaque materials			
High performance blind	0·70	0·30	0
Medium performance blind	0·55	0·45	0
Low performance blind	0·40	0·60	0
Translucent materials			
High performance blind	0·50	0·10	0·40
Medium performance blind	0·40	0·20	0·40
Low performance blind	0·30	0·30	0·40
Clear glass	0·07	0·13	0·80
Heat absorbing glass	0·05	0·51	0·44

The most important factor influencing the performance of blind systems is the reflectance and this is the basis of the designation of blind performance in Table I. A high performance translucent blind material has a reflectance of 0·50 or greater, a medium performance one has a reflectance of about 0·40, a low performance one has a reflectance of 0·30 or less. The values in Table I apply to the blind as a whole only when the material is flat and parallel to the plane of the window, that is when curtains are drawn together without folds or when louvers are completely closed. When louvers are partly open the solar optical characteristics are modified by inter-reflections and the effect can be assessed by computer routines that calculate the angle factors on which the interchange of radiant energy between adjacent louvers depends. In calculating the modified characteristics, which can be done for any angular setting of the louvers, it is assumed that the louvers are adjusted so that there is no direct sun penetration. A typical arrangement, especially with horizontal louvers, is to set them at 45° to the plane of the window and this condition was assumed in calculating the modified values shown in Table II for the materials listed in Table I. If there is to be no direct sun penetration, the angle of incidence of the solar radiation on the glass must be at least 22½°. The calculation assumes that the radiation is normal to the louvers and therefore incident at an angle of 45° to the glass. Table II shows the resulting optical properties of the glasses. **1** shows the relation between the basic properties of louver materials and the properties of a blind system with louvers set at 45° and no direct sun penetration.[3]

Table II. Solar optical properties of blinds with louvers set at 45°

Blind material	Reflectance	Absorptance	Transmittance
Opaque materials			
High performance blind	0·50	0·39	0·11
Medium performance blind	0·40	0·53	0·07
Low performance blind	0·30	0·65	0·05
Translucent materials			
High performance blind	0·50	0·14	0·36
Medium performance blind	0·40	0·27	0·33
Low performance blind	0·30	0·40	0·30
Clear glass	0·09	0·13	0·78
Heat absorbing glass	0·07	0·53	0·40

The values of the heat transfer coefficients at the glass surfaces, which affect the proportion of the absorbed energy that is released inwards, are derived from the surface resistances recommended in the *IHVE Guide*[4] for normal exposure, i.e. outside surface resistance = 0·055 m^2K/W, inside surface resistance = 0·123 m^2K/W. For double glazing with an internal blind the thermal resistance of a 12 mm air space is used (0·154 m^2K/W); for double glazing with blind between the thermal resistance of a 20 mm, or greater, air space is used (0·167 m^2K/W).

Shading coefficients have been computed for a range of blind materials including the various permutations of the typical materials listed in Table I. The details for these are shown in Table III. The shading coefficients listed for closed louvers apply also to curtains or roller blinds made of materials with the same basic properties as the louver materials.

2, 3, 4 and **5** relate the total shading coefficients of a variety of window systems to the transmittance of the glass and the reflectance of the blind. In the case of louvered blinds, the appropriate reflectance is that of the blind with louvers set at 45° and not the reflectance of the louver material. The translucent blind, **5**, is assumed to be of material with a transmittance of 0·40.

2 to **5** can be used with reasonable assurance for predicting the total shading coefficient. This single figure does not always give sufficient information and the short-wavelength and long-wavelength shading coefficients are needed, especi computer studies. The short-wavelength shading co defines the portion of the incident solar radiation that is

Table III. Shading coefficients

Window design	Blind performance	Opaque louver material Shading coefficient Short wave	Long wave*	Total	Translucent louver material Shading coefficient Short wave	Long wave*	Total
Single glazing without blind							
Clear glass	—	0·92	0·05	0·97	0·92	0·05	0·97
Heat absorbing glass	—	0·51	0·18	0·69	0·51	0·18	0·69
Single glazing with louvers closed							
Clear glass	High	0	0·36	0·36	0·38	0·16	0·54
,, ,,	Medium	0	0·50	0·50	0·38	0·25	0·63
,, ,,	Low	0	0·63	0·63	0·38	0·34	0·72
Heat absorbing glass	High	0	0·40	0·40	0·21	0·27	0·48
,, ,,	Medium	0	0·46	0·46	0·21	0·31	0·52
,, ,,	Low	0	0·52	0·52	0·20	0·36	0·56
Single glazing with louvers at 45°							
Clear glass	High	0·10	0·43	0·53	0·34	0·19	0·53
,, ,,	Medium	0·07	0·55	0·62	0·31	0·31	0·62
,, ,,	Low	0·05	0·65	0·70	0·28	0·42	0·70
Heat absorbing glass	High	0·05	0·41	0·46	0·17	0·29	0·46
,, ,, ,,	Medium	0·03	0·47	0·50	0·15	0·35	0·50
,, ,, ,,	Low	0·02	0·52	0·54	0·14	0·40	0·54
Double glazing without blind†							
Clear glass	—	0·74	0·10	0·84	0·74	0·10	0·84
Heat absorbing glass	—	0·41	0·14	0·55	0·41	0·14	0·55
Double glazing with louvers between, closed†							
Clear glass	High	0	0·15	0·15	0·32	0·10	0·42
,, ,,	Medium	0	0·21	0·21	0·31	0·14	0·45
,, ,,	Low	0	0·26	0·26	0·31	0·18	0·49
Heat absorbing glass	High	0	0·19	0·19	0·17	0·16	0·33
,, ,, ,,	Medium	0	0·21	0·21	0·17	0·17	0·34
,, ,, ,,	Low	0	0·23	0·23	0·17	0·19	0·36
Double glazing with louvers between, at 45°†							
Clear glass	High	0·09	0·19	0·28	0·28	0·12	0·40
,, ,,	Medium	0·05	0·24	0·29	0·25	0·16	0·41
,, ,,	Low	0·04	0·27	0·31	0·23	0·20	0·43
Heat absorbing glass	High	0·04	0·20	0·24	0·14	0·16	0·30
,, ,, ,,	Medium	0·03	0·22	0·25	0·13	0·18	0·31
,, ,, ,,	Low	0·02	0·23	0·25	0·11	0·20	0·31
Double glazing with internal louvers, closed†							
Clear glass	High	0	0·40	0·40	0·31	0·22	0·53
,, ,,	Medium	0	0·50	0·50	0·31	0·29	0·60
,, ,, ,,	Low	0	0·60	0·60	0·31	0·35	0·66
Heat absorbing glass	High	0	0·32	0·32	0·17	0·22	0·39
,, ,, ,,	Medium	0	0·37	0·37	0·17	0·25	0·42
,, ,, ,,	Low	0	0·42	0·42	0·17	0·28	0·45
Double glazing with internal louvers, at 45°†							
Clear glass	High	0·08	0·44	0·52	0·27	0·25	0·52
,, ,,	Medium	0·05	0·53	0·58	0·25	0·33	0·58
,, ,,	Low	0·04	0·60	0·64	0·22	0·42	0·64
Heat absorbing glass	High	0·04	0·32	0·36	0·14	0·22	0·36
,, ,, ,,	Medium	0·03	0·37	0·40	0·13	0·27	0·40
,, ,, ,,	Low	0·02	0·40	0·42	0·11	0·31	0·42

* Includes long wavelength radiation and convected heat.
† in all the double glazing systems the inner glass is clear.

transmitted through the window with no change of wavelength, that is, for clear glass, most of the visible part of the spectrum and some of the near infra-red. The long-wavelength shading coefficient defines the portion of the incident solar radiation that is absorbed in the window and then released to the indoor space by conduction, convection and long-wavelength radiation. The window absorbs most of the infra-red and ultra-violet radiation and, especially with tinted glasses, some of the visible.

The short-wavelength shading coefficient (SCSW) may be calculated from the following approximate formula which is accurate to 1·5 per cent for ordinary, practical materials:

$$SCSW = C \times T_g \times T_b$$

where C = a constant

T_g = transmittance of glass (outer glass of double glazing)

T_b = transmittance of blind.

The value of C depends upon the window system:

Single glazing with internal blind, $C = 1 \cdot 190$
Double glazing with blind between, $C = 0 \cdot 985$
Double glazing with internal blind, $C = 0 \cdot 980$

The long-wavelength shading coefficient can be calculated by subtracting the short-wavelength coefficient from the total coefficient found in **2** to **5**.

The short-wavelength and long-wavelength shading coefficients are important because they reveal the characteristics of the solar heat gain process. For example, from Table III, the addition of medium reflectance louvers to a single-glazed heat-absorbing glass reduces the total shading coefficient from 0·69 to 0·50, an improvement of 28 per cent. But the long-wavelength shading coefficient is increased from 0·18 to 0·35, which nearly doubles the immediately sensible heat load. Only the short-wavelength component of the solar heat gain is subject to the delaying and reducing effect of the thermal capacity of the structure.[5] Care is therefore needed in the application of total shading coefficients to the estimation of the capacity of air conditioning plant.

The time delay in the effect of the admitted short-wavelength radiation must be appreciated when the routine of adjustment of blinds is established. This means, for example, that when in spring and autumn office workers find that their east facing offices are hot when they arrive for work it is already too late to close the blinds because most of the effective heat was admitted two or three hours earlier. The situation could be improved by closing the blinds at night.

The most efficient operation of louvered blinds is achieved when they are so adjusted that the reflected radiation passes normally through the glass as shown in **6**. Because the position of the sun changes continually it is impracticable to keep louvers in the most efficient position and it is general practice to make an initial setting, probably to somewhere near the 45° position assumed in this chapter, and then leave the blind unchanged until the sun is off the facade. Because the apparent motion of the sun is predominantly vertical at sunrise and sunset and horizontal at noon, the practice of leaving blinds in one position favours the choice of vertical louvers for east and west facing windows and horizontal louvers for windows facing towards the equator.

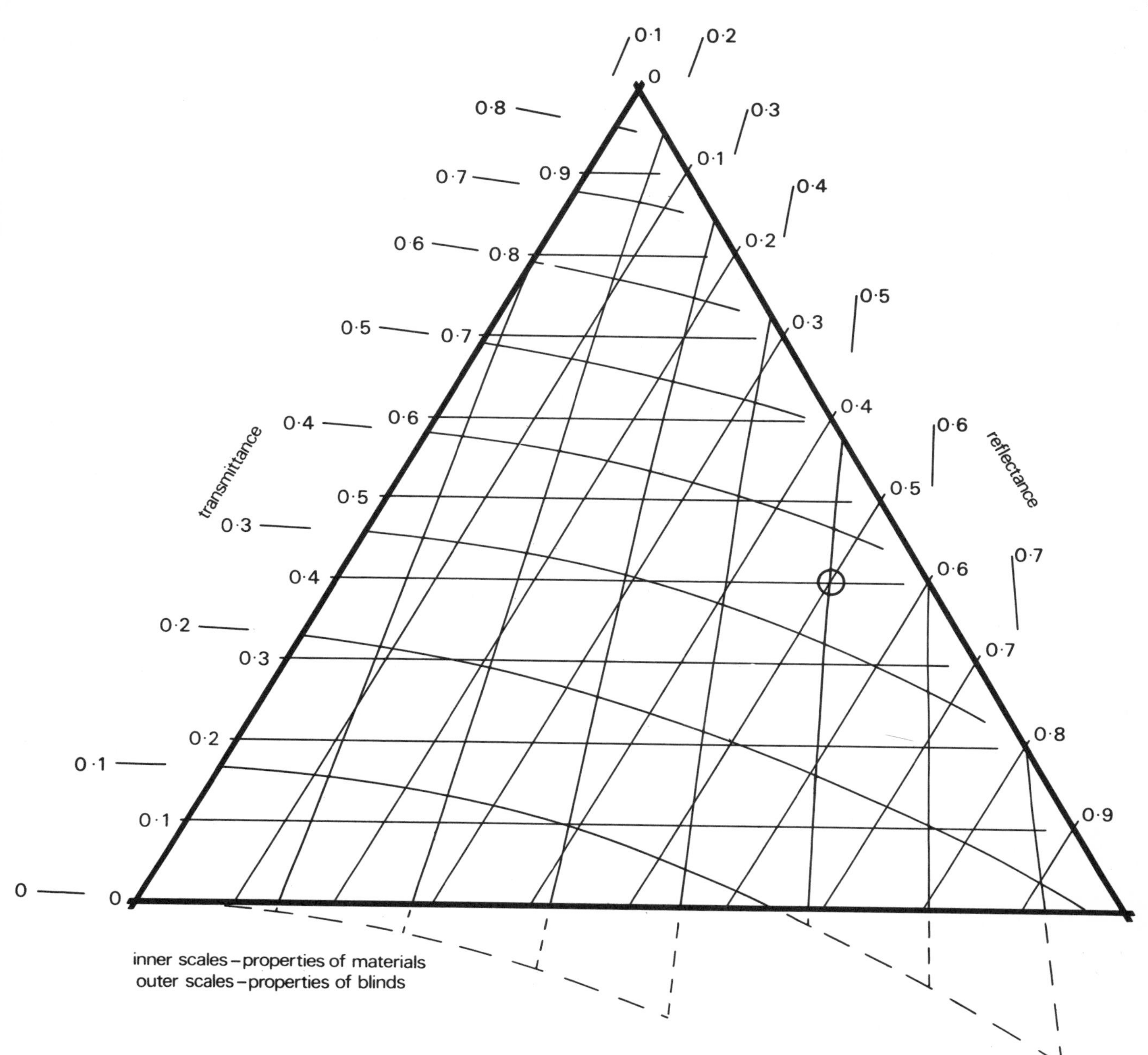

1 *The relation of properties of blind materials to the properties of louver blinds with the louvers set at 45° and no direct sun penetration. For example, a translucent high performance material (reflectance, 0·50; transmittance, 0·40) is plotted at the ringed point on the 'properties of materials' grid. The properties of the blind (reflectance, 0·50; transmittance, 0·36) are read off the 'properties of blinds' grid. The absorptance scales have been omitted to improve legibility. Absorptances are easily derived because the sum of reflectance, absorptance and transmittance is unity.*

The prediction of daylighting conditions is not relevant to the use of blinds. Daylight design is usually intended to ensure that sufficient daylight is available on dull, overcast days when there is no need to close the blinds because there is no excess solar radiation to control and no glare to combat.

One effect of high performance blinds is to reflect a large proportion of the transmitted energy for a second pass through the glazing so that the energy absorbed in the glass is increased and the temperature of the glass rises. With some glasses in some glazing systems this increased temperature can introduce a risk of thermal fracture of the glass. This effect is discussed in detail in Chapter 9.[6] The order of importance of the effect is exemplified in Table IV where the listed temperature differences are the increases in temperature of the central area of the glass above the temperature of the shielded edge for a peak intensity of solar radiation of 750 W/m^2, and a diurnal range of outdoor temperature of 10 K. The references in Table IV to 'borderline' conditions emphasise that simplified prediction systems must be based on the worst values of all the parameters that are assumed to be fixed. Real cases can be analysed in much more detail and those that give a borderline result by the simplified system of Chapter 9 should be referred to the manufacturer for assessment.

Properties of blind materials

The prediction of shading coefficients for windows with blinds, as described in this chapter, depends upon a knowledge of the solar optical properties of the material of the blind. In some cases the necessary information will be available from the manufacturer of the blind but, at present, it is more likely that the manufacturer will not know.

The required properties are the reflectance, absorptance and transmittance of the material to radiation of the spectral distribution defined by Parry Moon's curve for air mass 2, that is for the radiation from the sun when it is at an altitude of 30°. In the laboratory, it is usual to measure reflectance and transmittance at each frequency throughout the spectrum (an

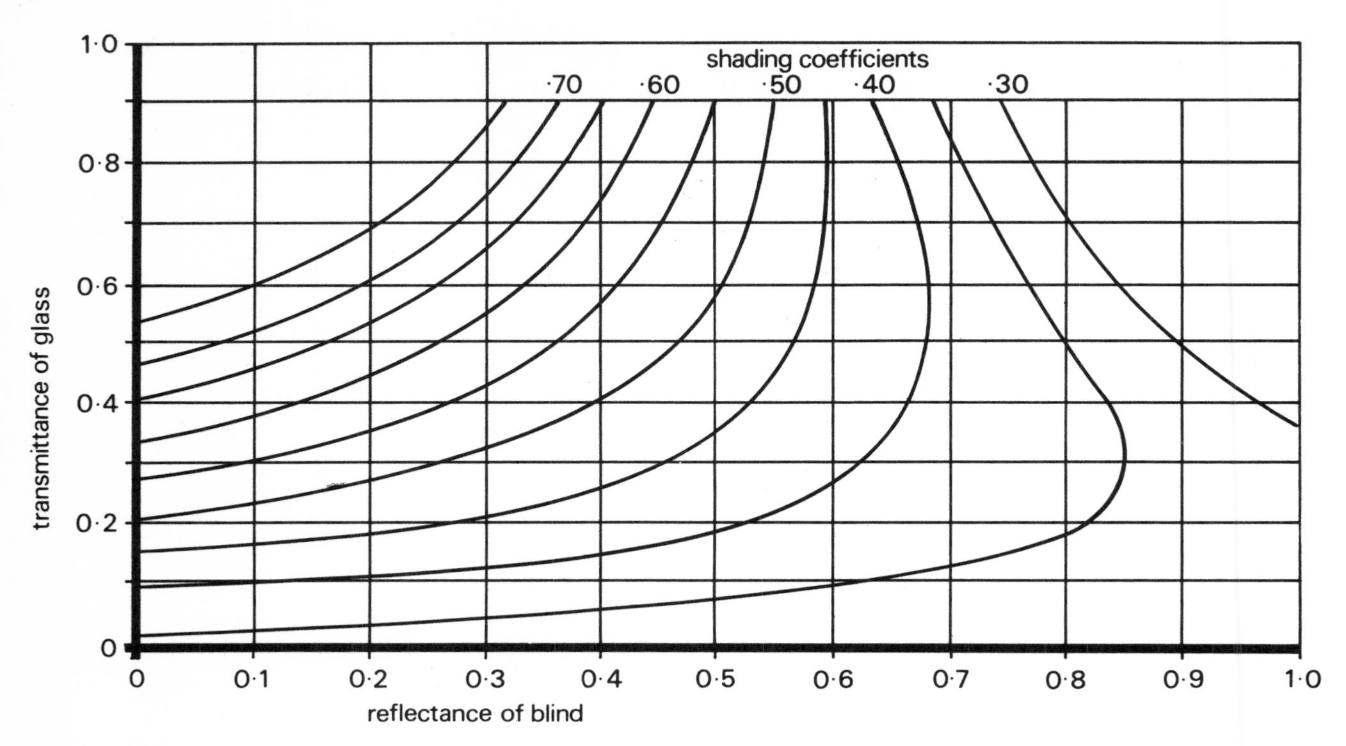

2 *The shading coefficient of SINGLE GLAZING with a blind of opaque material.*

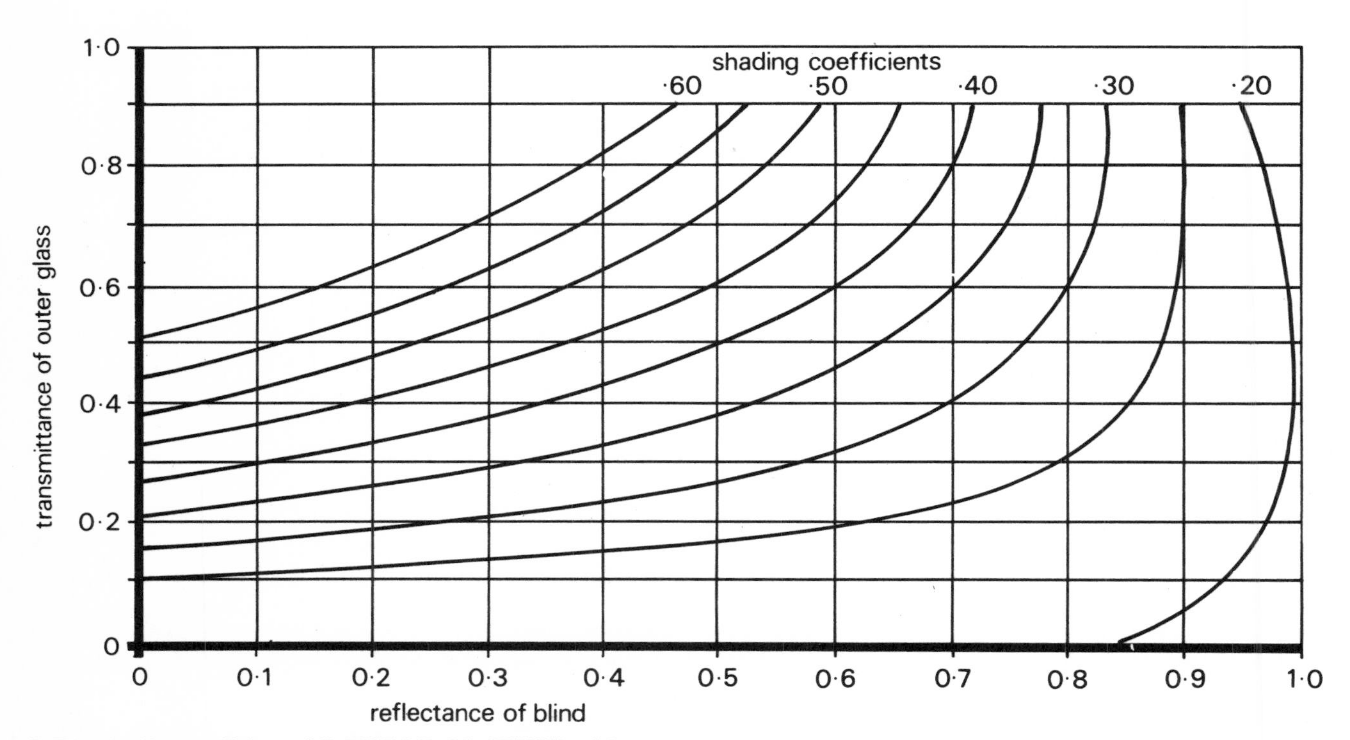

3 *The shading coefficient of DOUBLE GLAZING with an INTERNAL BLIND.*

Table IV. Safety from thermal fracture.

Window design	Blind performance	Temperature difference (K)	Remarks
Single glazing, internal blind			
Clear glass	High	15	Safe
,, ,,	Low	13	Safe
Heat absorbing glass	High	32	Safe
,, ,, ,,	Low	30	Safe
Double glazing (outer glass) internal blind			
Clear glass	High	20	Safe
,, ,,	Low	16	Safe
Heat absorbing glass	High	41	Safe in metal frame, borderline in wood
,, ,, ,,	Low	37	Safe in wood frame, borderline in concrete
Double glazing (outer glass) blind between			
Clear glass	High	28	Safe
,, ,,	Low	25	Safe
Heat absorbing glass	High	49	Safe in gasket glazing
,, ,, ,,	Low	46	Safe in gaskets, borderline in metal

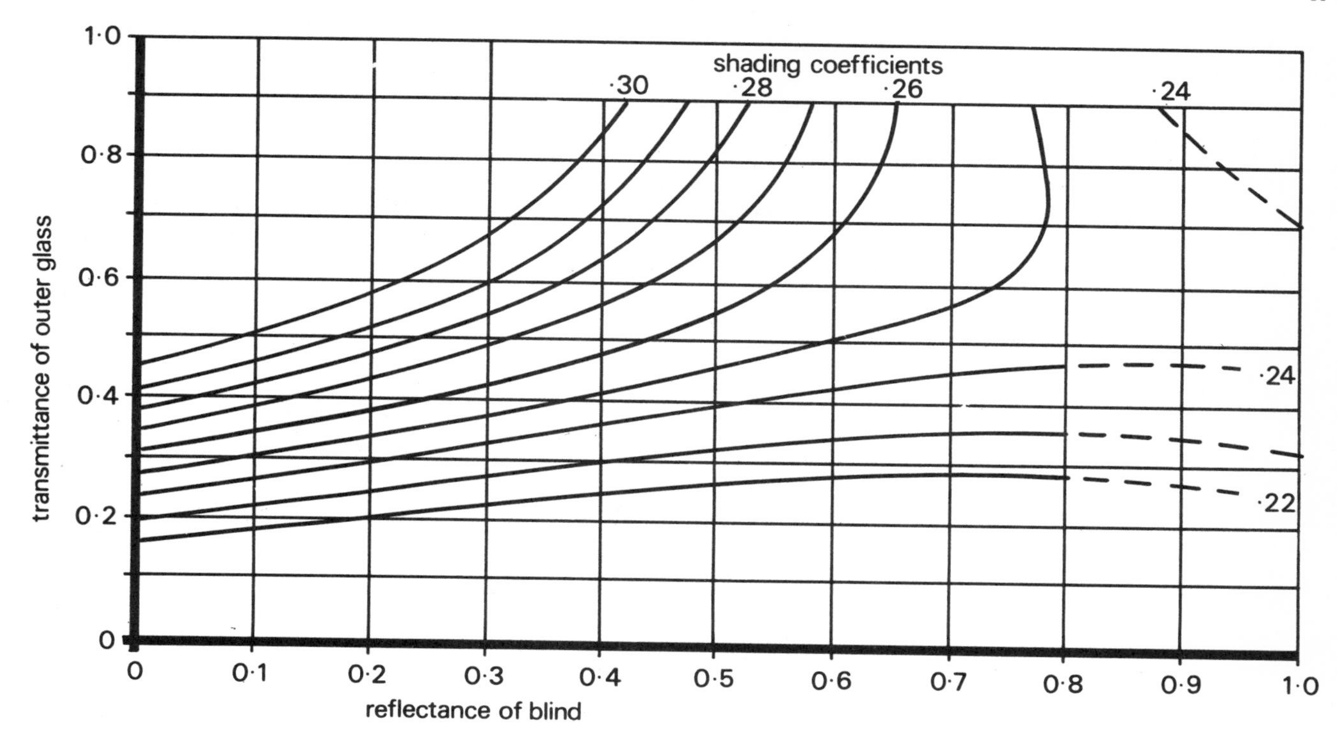

4 *The shading coefficient of DOUBLE GLAZING with an OPAQUE BLIND BETWEEN.*

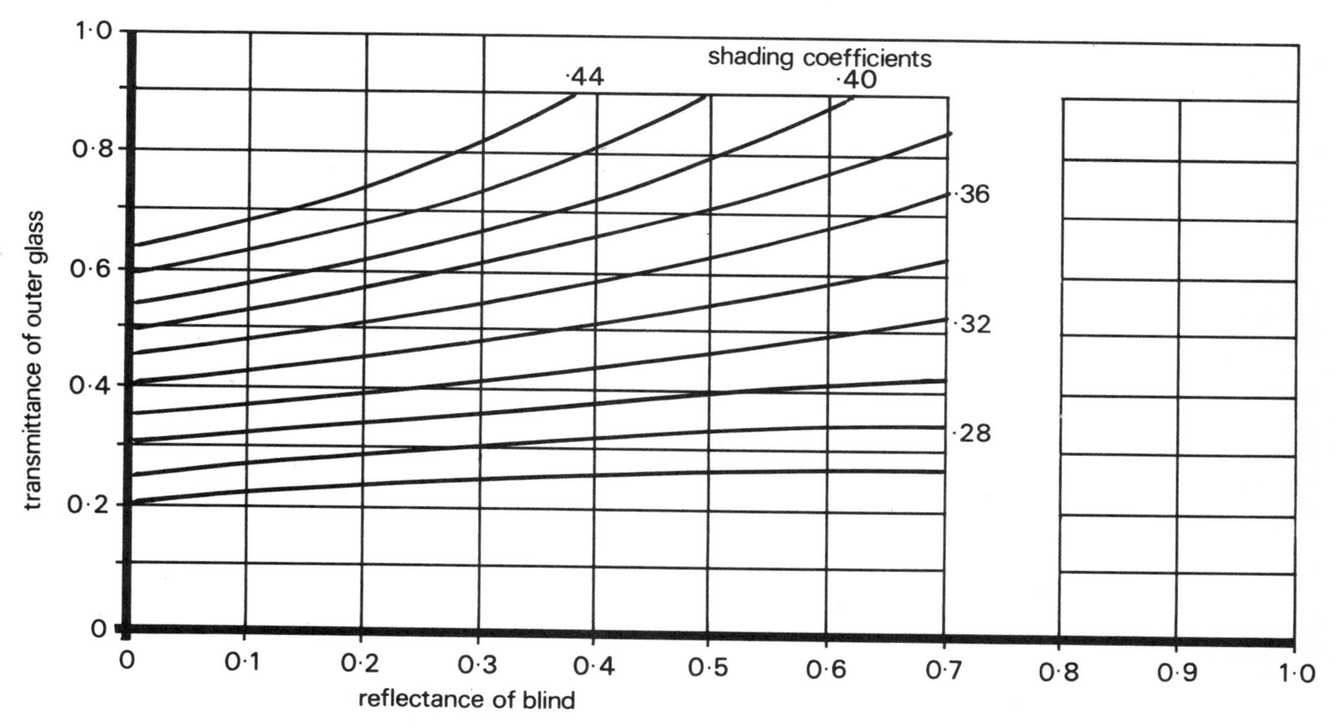

5 *The shading coefficient of DOUBLE GLAZING with a TRANSLUCENT BLIND BETWEEN. The material has a transmittance of* 0·40.

artificial source can be used for this) and to calculate the values for the solar spectrum. The absorptance is obtained by difference because any energy not reflected or transmitted is absorbed. The laboratory method therefore needs sophisticated spectrophotometric equipment and access to a computer to speed up the conversion calculations.

In the absence of such facilities some method of estimating the properties to an acceptable order of accuracy is required and this is more easily achieved if it is assumed that the properties as measured in sunshine over the whole solar spectrum are not greatly different from the properties as seen by the eye for just the visible part of the spectrum. This assumption will generally give sufficiently accurate results, although there is always a chance that an unfamiliar material may have properties by infra-red radiation very different from its properties by visible light. A method based on this assumption is described in the *ASHRAE Handbook of Fundamentals*.[7] An *openness factor*, the ratio of the open area between the fibres to the total area of the fabric, can be measured and used to classify fabrics but the classes may also be assessed by eye. With *closed* fabrics (open-

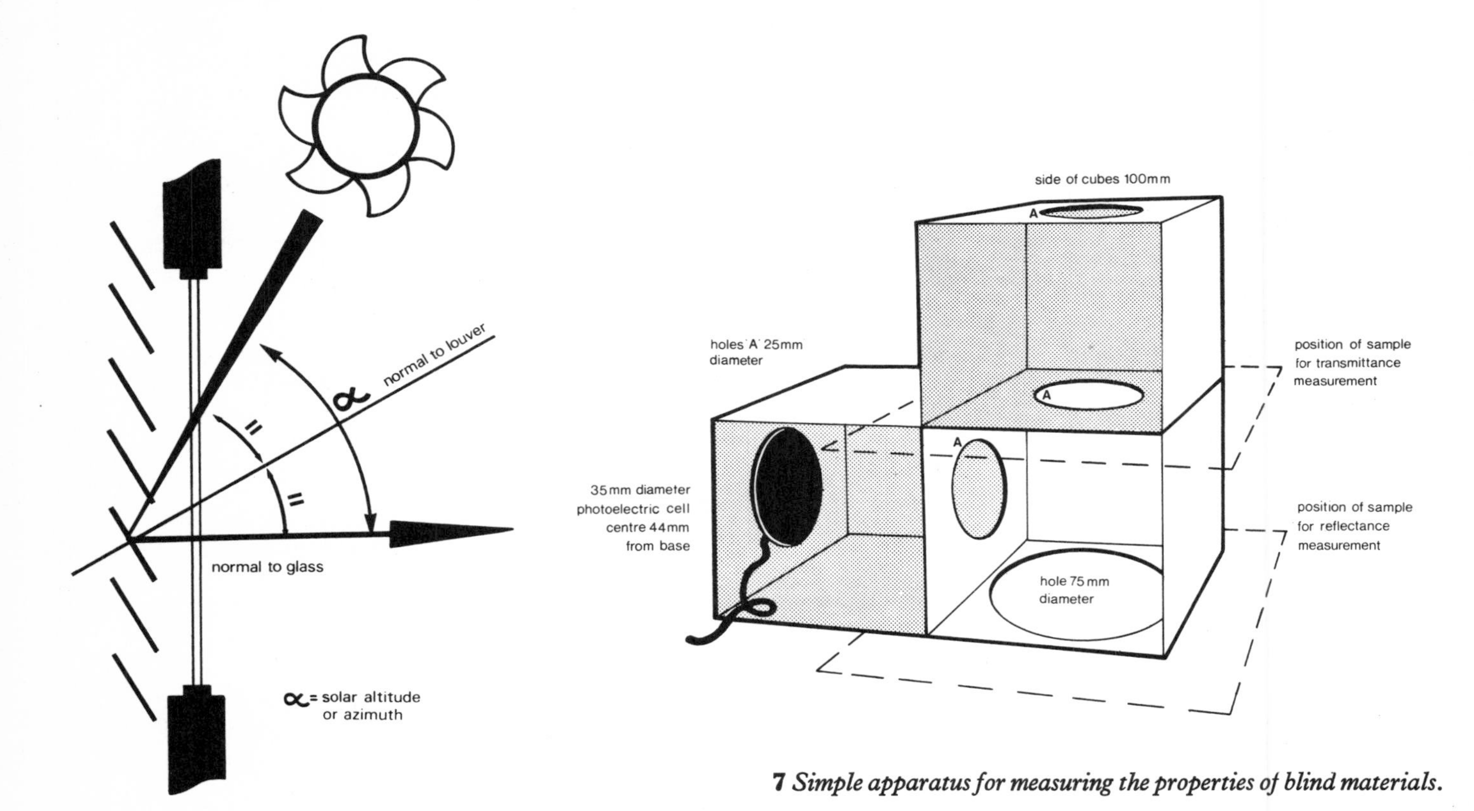

7 *Simple apparatus for measuring the properties of blind materials.*

6 *Louvered blinds adjusted for maximum rejection of solar radiation.*

ness factor = 0·00-0·07) no objects are visible through the material, and large light or dark areas may perhaps show. *Semi-open* fabrics (openness factor = 0·07-0·25) do not permit details to be seen and large objects are clearly defined. *Open* fabrics (openness factor = 0·25 and over) allow details to be seen and the general view is relatively clear with no confusion of vision. *Light*, *medium* and *dark* fabrics may be identified by eye, keeping in mind that it is the yarn colour or shade of light or dark that is being observed. These observations are used to select the appropriate fabric properties from Table V. The

Table V. Estimated properties of blind materials

Fabric description	Transmittance	Reflectance
Open, dark	0·39	0·07
Open, medium	0·49	0·25
Open, light	0·58	0·36
Semi-open, dark	0·18	0·10
Semi-open, medium	0·29	0·32
Semi-open, light	0·41	0·48
Closed, dark	0·05	0·14
Closed, medium	0·11	0·38
Closed, light	0·17	0·63

reflectance of a material can be estimated by matching the colour to a reference and using the Munsell system of colour coding.[8] For example, if the slats of a venetian blind match the pale blue shade 18 B 17 on Card 2 of BS4800,[9] the Munsell reference is given as 5 B 8/1 of which the digit 8 is the Munsell value (v). The reflectance (r) can be calculated from $r = v(v-1)\% = 8 \times 7\% = 56\% = 0{\cdot}56$.

A simple apparatus that can be used to measure the transmittance and reflectance of blind materials is shown in **7**. It consists of three cardboard cubes of 100 mm side, two of them black inside and one white, with circular holes cut as shown. A lamp (a 250 W spot lamp is best) is shone into the top cube to illuminate the sample and the luminance of the white integrating cube is measured by the photoelectric cell connected to an inexpensive microammeter with a full scale deflection of about 10 microamps.

To measure reflectance, the sample is placed in the lower position with a black surface underneath it and the microammeter reading is noted. The sample is then replaced by a white reference surface of known reflectance and the new meter reading is taken. The ratio of the reflectances is the same as the ratio of the meter readings. If no better reflectance standard is available, clean white blotting paper may be used: it has a remarkably consistent reflectance of 0·85.

To measure transmittance, a white surface is placed in the lower sample position and the meter readings are compared with the sample in and out of the upper position. The ratio of the readings gives the transmittance.

References

1 *Thermal transmission of windows*. Environmental Advisory Service Report No 1. Pilkington Brothers Limited, March 1973. Third edition.
2 Moon, P. 'Proposed standard radiation curves for engineering use'. *Journal of the Franklin Institute*. Vol 230, November 1940.
3 The expressions for transmitted and reflected energy are derived as equations 5 and 6 in Appendix 1 to *Vertika blinds, solar performance data*, published by Vertical Blinds Limited, Brook St, Macclesfield, 1972. Figure **1** is based on these equations.
4 The Institution of Heating and Ventilating Engineers. *IHVE Guide, Book A*. London 1970.
5 *Solar heat gain through windows*. Environmental Advisory Service Report No 2, Third edition. Pilkington Brothers Limited, August 1972.
6 'The application of solar control glasses'. *Glass and Windows Bulletin No* 10. Pilkington Brothers Limited, January 1972.
7 American Society of Heating, Refrigerating and Air-Conditioning Engineers. *Handbook of Fundamentals*, Chapter 28. New York, 1967.
8 *Windows and environment*. Pilkington Brothers Limited, 1969. Part II, section 17.
9 British Standards Institution. BS4800: 1972. *Paint colours for building purposes*.

Part III Solar control

9 Thermal safety

The possibility of the fracture of glass under thermal stresses was mentioned in Chapters 1 and 8. This chapter describes the effect in more detail and shows how the factors on which thermal safety depends can be calculated.

Thermal stress in glass

The cause of thermal stress

Glass is normally fixed in buildings by retaining the edges of the glass behind a glazing bead or a fillet of glazing compound. In sunshine, the central area of the glass exposed to solar radiation absorbs heat, rises in temperature and expands, **1**. The edges, shielded from the radiation by the edge cover, remain cooler and the resulting differential expansion causes stresses which, if they reach the breaking stress of the glass at any point, will cause a thermal fracture, **2**. A rapid rise in the air temperature on either side of the glass has a similar effect. The magnitude of the stress in a given situation depends on the temperature difference between the hottest and coldest areas of the glass and also on the detailed distribution of the temperature gradients across the glass. Any factor that encourages the hot-centre/cold-edge condition tends to increase the stress. High levels of solar radiation, large absorptances, and indoor blinds that reflect radiation back through the glass all increase the heat absorbed by the exposed areas of glass. High air temperatures, low rates of air movement, and the insulation provided by blinds and multiple glazing tend to reduce the loss of heat and uphold the centre temperature. Low temperatures at the edges are maintained by easy paths of conduction from the glass through the frame to a cold building structure with a large thermal inertia.

The effect of climate

The intensity of solar radiation is determined by the latitude, altitude and orientation of the building, the season and time of day, the amount of cloud cover and atmospheric pollution, and by reflections from the ground or adjacent structures. The temperature and movement of air within the building will depend upon the type and disposition of the heating and cooling equipment and on the control system used but there will be a tendency for the indoor temperature to be maintained at a constant level. The outdoor air temperature may vary considerably from a minimum near dawn to a maximum at about 1400 to 1500h. The exposed areas of glass tend to follow these changes but, because of the effect of the thermal inertia of the structure, the shaded edges tend to remain at the dawn temperature and increase in temperature at a slower rate than the exposed areas so that the temperature difference is increased, **3**. This factor is taken into account by the air temperature variation map, **6**.

The effect of the type of glass

The distribution of the incident solar radiation by the three

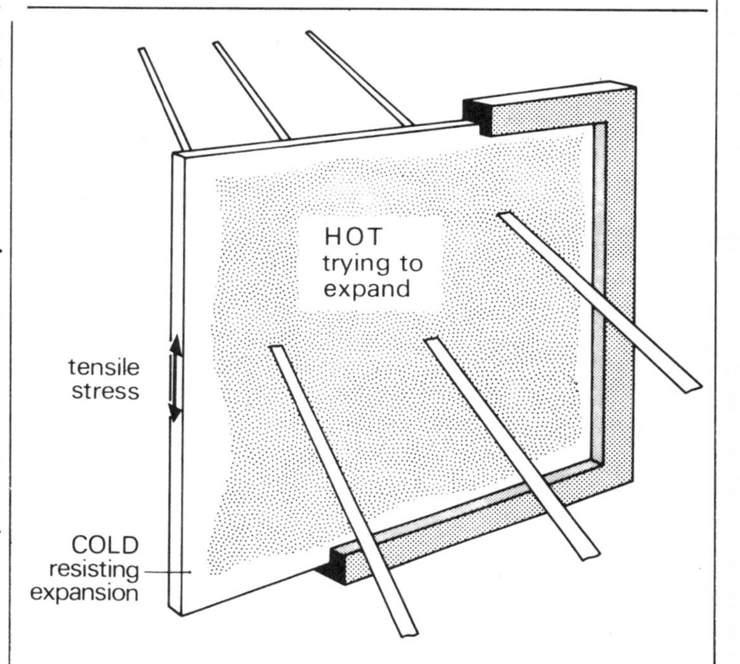

1 *The generation of thermal stress in glass.*

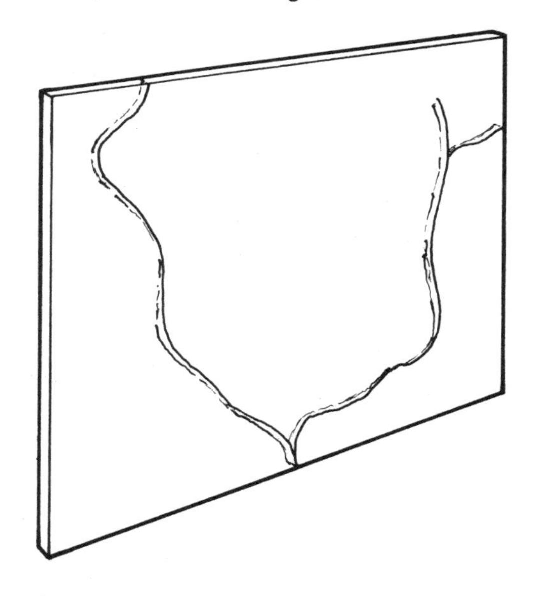

2 *A typical thermal fracture.*

processes of reflection, absorption and transmission, as discussed in paragraph **7.03** of Chapter 1, depends upon the properties of the selected glass. The relevant properties of glasses and double glazing combinations are shown in Table XII, Function 8, Chapter 1 to illustrate the range that is

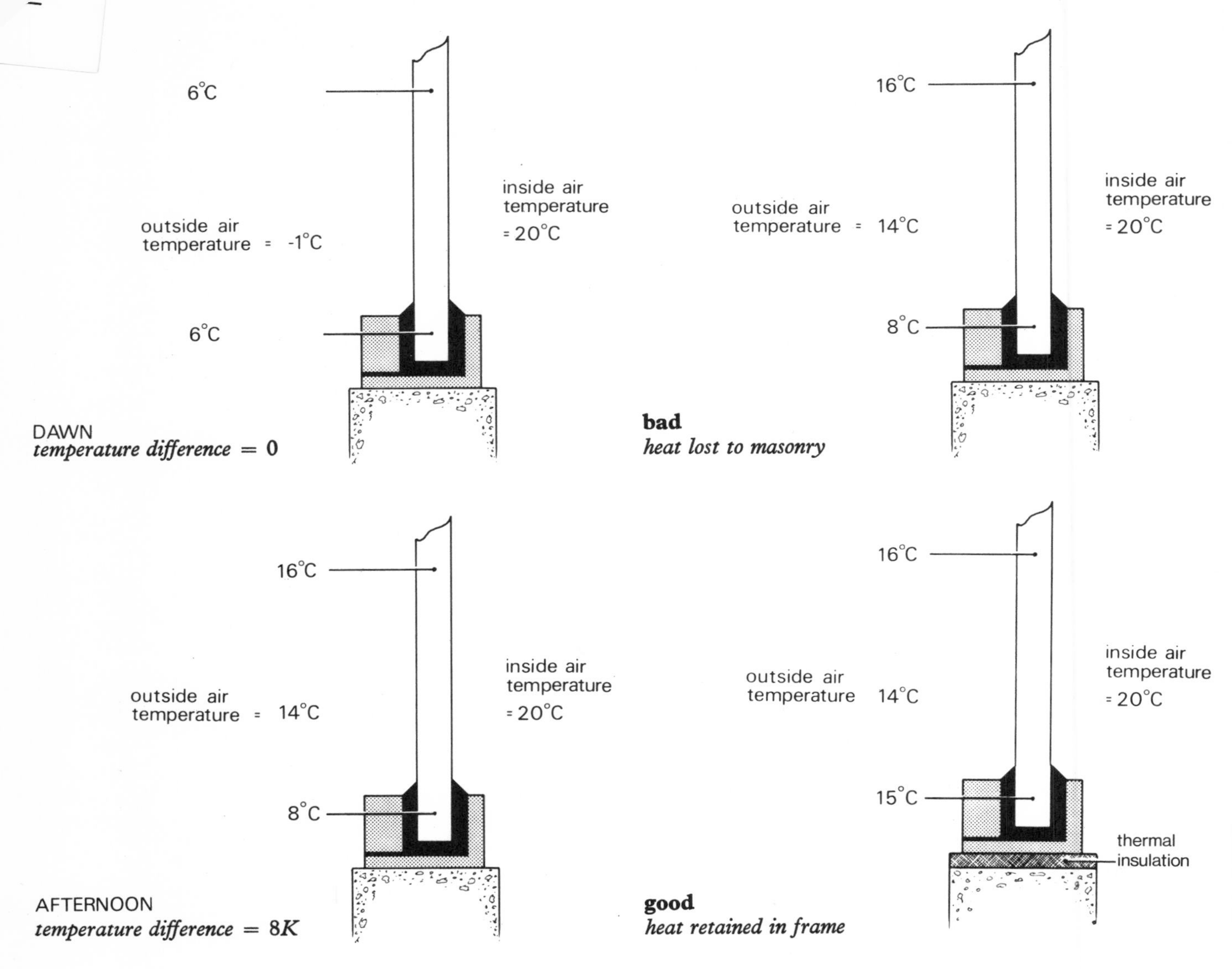

3 *Loss of heat to the structure tends to keep the temperature of the edge of the glass down to the dawn value. Any solar radiation will increase the temperature difference between the edge and the central areas.*

4 *Insulating the frame from the structure allows the temperature of the edge of the glass to rise. Any solar radiation will increase the temperature difference between the edge and the central areas.*

available but it will probably be necessary to obtain the data for a particular glass from the manufacturer.

The effect of the glass edge
Because the normal mode of thermal breakage of glass is by the action of tensile stress located in and parallel to an edge, the breaking stress of the glass is mainly dependent on the extent and position of flaws in the edges. The condition of the glass edge is, therefore, extremely important and the prediction technique described below applies to glass with good, undamaged edges, glazed in accordance with recommended practice (see 'Glazing solar control glasses', below).
The size of the panes affects the safety of the glass in two ways. First, the larger and thicker squares are more difficult to cut, handle and glaze without causing damage to the edges and introducing dangerous flaws. Second, the probability of a critical flaw being present in the edge is reduced for smaller values of the area of the edge (the product of perimeter and thickness). To take account of this second effect, **10** allows an increased design stress for edge areas less than 30 000 mm^2, that is, 1·25 m square for 6 mm thick glass. In sealed double glazing units the edge area to be taken is that of each glass separately.

The effect of the glazing system
The temperature gradients and, therefore, the thermal stresses in the glass also depend on the details of the window design. The selected glass determines the amount of solar energy absorbed: this is usually the major factor. Multiple glazing systems have temperature distribution patterns different from those of single glazing because each glass, having absorbed its share of solar energy, transfers some of its heat to other glasses in the system.
The materials used in framing the glass vary widely in their thermal properties and the details of the frame design are, therefore, important in determining the temperature gradients near the edges of the glass. In general, if the frame is insulated from the surrounding building fabric it will tend to warm up at the same rate as the glass, **4**. This applies whether the heating is due to solar radiation or to rising air temperatures. If the frame is in good thermal contact with a heavy masonry structure it will lose heat rapidly to the masonry and stay cool.
Where a metal frame incorporates a region of high thermal resistance, design to reduce the heat transfer through the frame, the resistive region must be to the outside of the line of the glazing. This allows the temperature of the frame near the edge of the glass to rise but a thermal resistance to the inside en-

courages the undesirable cold-edge condition in the glass. Blinds and other shading devices interfere with the free movement of air over the glass and they reflect, absorb and re-radiate solar radiation. All these have effects on the temperature of the glass that must be taken into account.
Edge cover is instrumental in causing stress in the edge of sunlit glass but, within the practicable limits set by other considerations, changes in the width of edge cover alter the edge stress so little that their effect on the thermal safety of the glass is negligible.

Assessing thermal safety

This section presents a method by which the safety of a solar control glass can be assessed in a situation where thermal stresses are liable to be generated. The method applies to glass with good, undamaged edges, glazed in accordance with recommended practice (as described in Chapter 1 and 'Glazing solar control glasses' below) and exposed to sunshine, including ground reflections but with no additional source of artificial radiation. It depends on:
(i) calculation of the maximum service stress to which the glazing system will be subjected.
(ii) calculation of the design stress of the glazing system.
(iii) comparison of the calculated values, to test if the design stress is likely to be exceeded in service.

(i) Calculating the maximum service stress

(a) From maps **5** and **6** find the values of the intensity of solar radiation on vertical surfaces and the diurnal range of air temperature that apply to the location. Use these values to find, in the appropriate column of Table I, II or III, the basic temperature difference for the proposed glazing system.
In the tables, the glasses are identified by type, thickness and colour, and by two numbers: the first is the visible transmittance and the second is the total transmittance (expressed as percentages). Thus 'body-tinted 6 mm 50/60 (bronze)' describes a 6 mm thick glass that is coloured bronze throughout, has a visible transmittance of 0·50 and a total transmittance of 0·60—it is the third glass tabulated in Table XII, Function 8, Chapter 1.
If the basic temperature difference is tabulated in a shaded area of the table the glazing system can be accepted as safe for the proposed location without continuing the calculation.
For double glazing with a ventilated air space, that is, having an air flow equivalent to a wind speed of 8 km/h, the basic temperature difference should be reduced by 2K.
When more detailed information of local radiation and temperature conditions is available, as in the radiation map of Great Britain, **7**, it may be used to improve the estimate of basic temperature difference. The two parameters needed to find the right column of Table I, II or III are:
Solar radiation intensity—the maximum intensity of solar radiation on a vertical surface. This maximum intensity will exceed that usually adopted for heat gain calculations but it may be modified by the orientation of the glazing. For example, in temperate latitudes glass that faces the nearer pole cannot be subjected to peak radiation intensities.
Diurnal temperature range—the maximum of the average daily temperature range taken over a period of several years.
(b) Correct the basic temperature difference for the effect of any blind or curtains by adding the appropriate correction from Table IV. For the purpose of this correction, a ventilated air space is one created by a blind 50 mm or more from the glass and with a gap of at least 50 mm at the top and bottom of the blind.
(c) Use the corrected temperature difference to find, from the graph, **8**, the maximum working stress.
(d) Multiply the maximum working stress by the appropriate factor from Table V to correct for the effect of the framing material and so obtain the maximum service stress. Where a combination of framing materials is used, the frame factor is determined by the material that is effectively in contact with the glass and also directly exposed to solar radiation, **9**.

(ii) Calculating the design stress

Calculate the area of edge of the proposed square of glass by multiplying the perimeter by the glass thickness, both quantities in millimetres. Use this area to find, from the graph, **10**, the design stress for the glass.

(iii) Comparing service and design stresses

If, on comparison, the service stress is less than or equal to the design stress, the glazing system can be accepted as safe for use in the proposed location, provided that the edges of the glass used are of an acceptable quality.
If the service stress exceeds the design stress the risk of thermal fracture is too high for the glazing system to be accepted in its proposed form. Inspection of the contributing factors in the calculation shows what changes may be made to reduce the service stress to an acceptable level. This is discussed further in 'Recommendations', below.

Example

It is proposed to glaze the eastern facade of a building in Liverpool with 12 mm body-tinted Float 18/44 (grey) in squares 1750×2000 mm. The glazing will be in wooden frames and light-coloured curtains will hang 70 mm from the glass with at least 60 mm clearance at top and bottom. Is the glass safe from thermal fracture?
From **6** and **7,** Liverpool has solar radiation at 750W/m^2 and a diurnal temperature range of 10K. Table I shows that, for these climatic factors, 12 mm body-tinted Float 18/44 (grey) has a basic temperature difference of 36K. Because this value is not found in a shaded area of the table the calculation must proceed.

Basic temperature difference	= 36K
Correction for single glazing, light-coloured blind, glass-blind space ventilated, Table IV	= 5K
Corrected temperature difference (36+5)	= 41K
Maximum working stress, **8**	= $24 \cdot 0 \times 10^6$ N/m^2
Frame factor, wooden frame, Table V	= 0·9
Service stress ($0 \cdot 9 \times 24 \times 10^6$)	= $21 \cdot 6 \times 10^6$ N/m^2
Area of glass edge ($2 \times 3750 \times 12$)	= 90×10^3 mm^2
Design stress, **10**	= $21 \cdot 0 \times 10^6$ N/m^2

In this case the service stress ($21 \cdot 6 \times 10^6$ N/m^2) exceeds the design stress ($21 \cdot 0 \times 10^6$ N/m^2) and the glass would not be safe from thermal fracture.
The situation could be improved by, for example, changing to glazing in metal frames. This would change the frame factor to 0·8 and reduce the service stress to ($0 \cdot 8 \times 24 \times 10^6$) = $19 \cdot 2 \times 10^6$ N/m^2 which is less than the design stress.

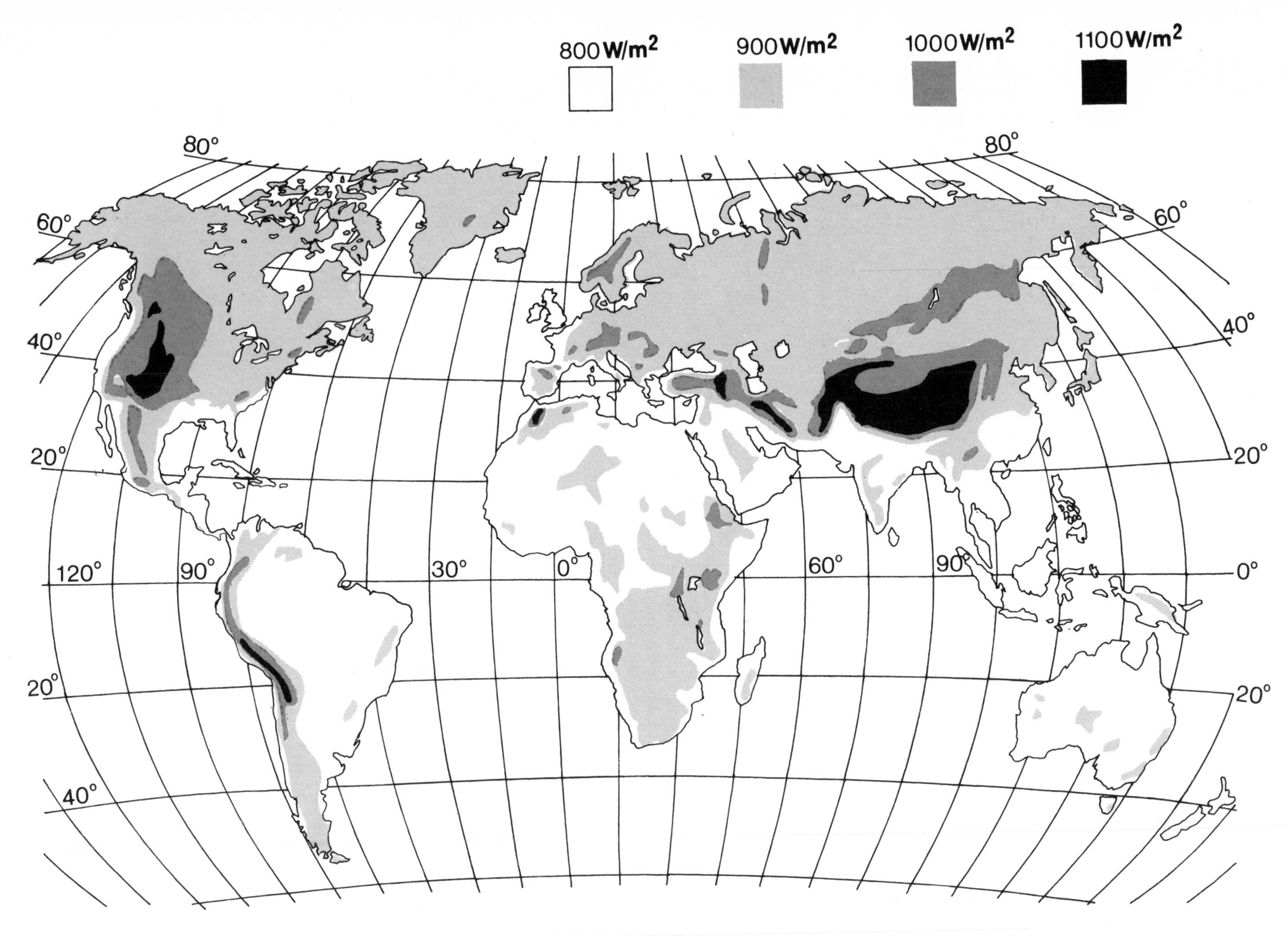

5 *Solar radiation intensities.*

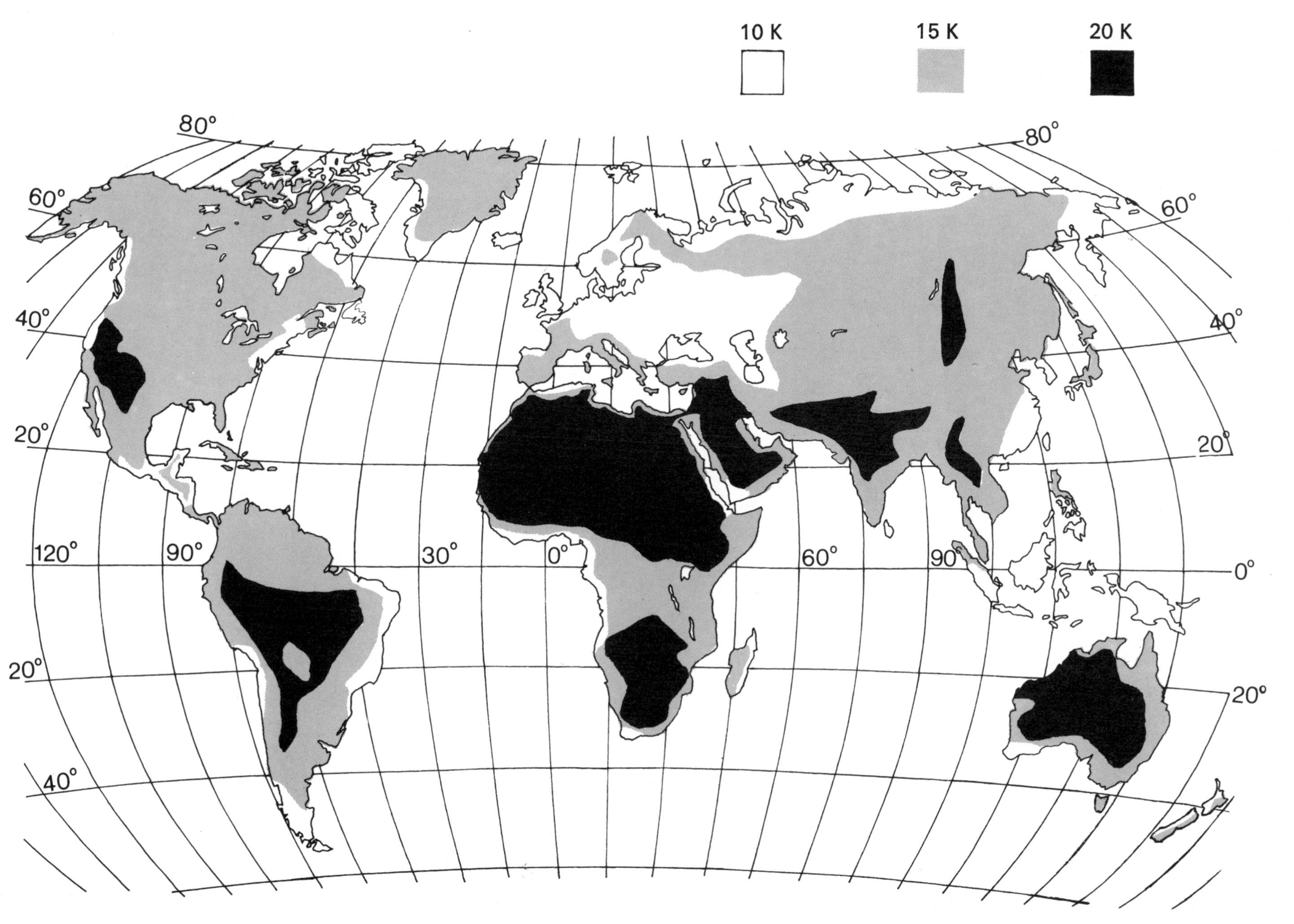

6 *Diurnal ranges of air temperatures.*

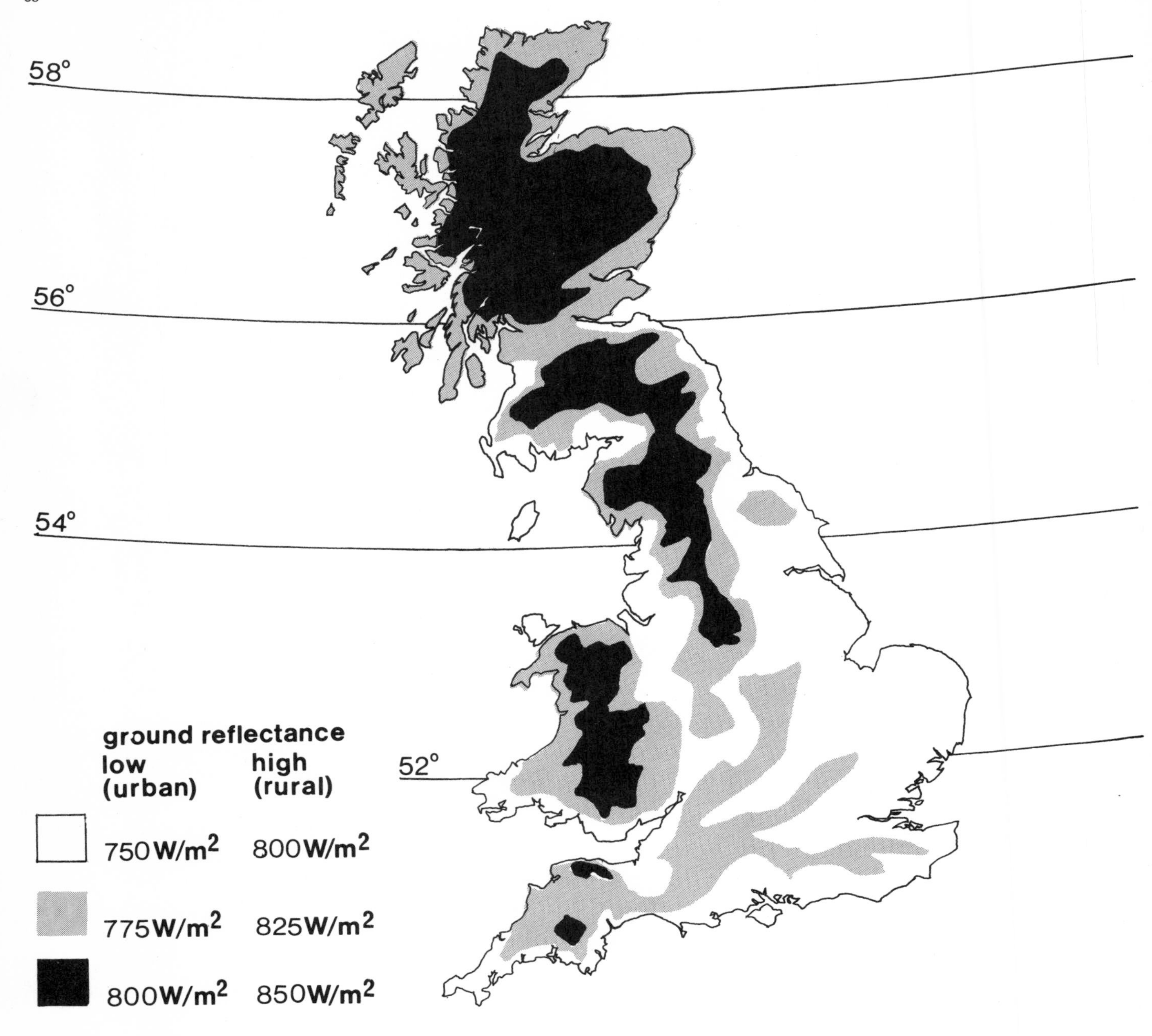

7 *Solar radiation intensities—Great Britain.*

Table I. Basic temperature differences—single glazing.
The shaded areas indicate, without further calculation, that the glass is safe from thermal fracture.

	Solar radiation intensity (W/m²) 750			775			800			825			850			900			1000			1100		
Diurnal temperature range (K)	10	15	20	10	15	20	10	15	20	10	15	20	10	15	20	10	15	20	10	15	20	10	15	20
6, 10 or 12 mm clear float	10	13	16	10	13	16	10	14	16	10	14	17	11	14	17	11	14	17	12	15	18	12	16	18
Surface-modified float																								
6 mm 49/66 (bronze)	19	22	25	20	23	26	20	23	26	21	24	27	21	24	27	22	25	28	24	27	30	26	29	32
10 or 12 mm 46/61 (bronze)	23	26	29	23	27	30	24	27	30	25	28	31	25	29	31	26	30	33	29	32	35	31	35	37
Body-tinted float																								
5 mm 78/65 (green)	22	26	29	23	26	29	24	27	30	24	28	30	25	28	31	26	29	32	28	32	34	31	34	37
6 mm 75/60 (green)	25	29	31	26	29	32	27	30	33	27	31	33	28	31	34	29	33	36	32	35	38	35	38	41
10 mm 66/50 (green)	32	35	38	33	36	39	33	37	39	34	37	40	35	38	41	37	40	43	41	44	47	44	47	50
6 mm 41/60 (grey)	27	29	32	27	30	33	28	31	34	29	32	34	30	32	35	31	34	36	34	37	39	37	39	42
10 mm 24/48 (grey)	34	37	39	35	38	40	36	38	41	37	39	42	38	40	43	39	42	45	43	46	49	47	50	53
12 mm 18/44 (grey)	36	39	42	37	40	43	38	41	44	39	42	45	40	43	46	42	45	48	47	49	52	51	53	56
6 mm 50/60 (bronze)	27	29	32	27	30	33	28	31	34	29	32	34	30	32	35	31	34	36	34	37	39	37	39	42
10 mm 32/48 (bronze)	34	37	39	35	38	40	36	38	41	37	39	42	38	40	43	39	42	45	43	46	49	47	50	53
12 mm 26/44 (bronze)	36	39	42	37	40	43	38	41	44	39	42	45	40	43	46	42	45	48	47	49	52	51	53	56
Body-tinted rough-cast																								
5 mm 55/48 (blue/green)	35	37	40	36	38	41	37	39	42	38	40	43	39	41	44	41	43	46	44	47	50	48	51	54
6 mm 48/45 (blue/green)	37	40	42	38	41	43	39	42	44	40	43	45	41	44	47	43	46	49	47	50	53	51	54	57
Reflective laminated glass																								
6 mm 17/24 (gold)	25	28	31	25	28	31	26	29	32	26	30	33	27	30	33	28	32	35	31	34	37	34	37	40
6 mm 16/22 (gold)	23	26	29	23	27	30	24	27	30	25	28	31	25	29	31	26	30	33	29	32	35	31	35	37
6 mm 39/34 (gold)	20	23	26	20	23	26	21	24	27	21	24	27	21	25	28	22	26	29	24	28	30	26	30	32

Table II. Basic temperature differences—OUTER GLASS of double glazing.
The shaded areas indicate, without further calculation, that the glass is safe from thermal fracture.

	Solar radiation intensity (W/m²) 750			775			800			825			850			900			1000			1100		
Diurnal temperature range (K)	10	15	20	10	15	20	10	15	20	10	15	20	10	15	20	10	15	20	10	15	20	10	15	20
6 mm clear float + 6 mm clear	16	20	24	16	21	24	17	21	24	17	21	25	17	21	25	18	22	26	19	23	27	20	25	28
Surface-modified float combinations																								
6 mm (bronze) + 6 mm clear	27	32	35	28	32	36	29	33	36	29	34	37	30	34	38	32	36	39	34	39	42	37	41	45
6 mm (bronze) + 6 mm (bronze)	31	35	38	31	36	39	32	36	40	33	37	41	34	38	42	35	40	43	39	43	46	42	46	50
10 or 12 mm (bronze) + 6 mm clear	32	36	40	33	37	40	34	38	41	34	39	42	35	39	43	36	41	45	40	45	48	44	48	52
10 mm (bronze) + 10 mm clear	33	37	40	34	38	41	34	39	42	35	39	43	36	40	44	38	42	45	41	46	49	45	49	53
12 mm (bronze) + 12 mm clear	33	37	41	34	38	42	35	39	43	36	40	44	37	41	45	39	43	46	42	46	50	46	50	53
Body-tinted float combination																								
6 mm (green) + 6 mm clear	35	39	42	36	40	43	37	41	44	38	42	45	38	43	46	40	45	48	44	48	52	48	52	56
6 mm (green) + 6 mm (green)	39	43	47	40	44	48	41	45	49	42	46	50	43	47	51	45	50	53	50	54	57	54	58	62
10 mm (green) + 6 mm clear	42	47	50	44	48	51	45	49	53	46	50	54	47	51	55	50	54	57	54	59	62	59	63	67
10 mm (green) + 10 mm clear	43	47	51	44	48	52	45	50	53	47	51	54	48	52	56	50	54	58	55	59	63	60	64	68
6 mm (grey) + 6 mm clear	37	40	44	38	41	45	39	42	46	40	43	47	41	44	48	43	46	50	46	50	54	50	54	57
6 mm (grey) + 6 mm (grey)	41	45	48	42	46	49	43	47	50	44	48	51	46	49	53	48	51	55	52	56	59	57	60	64
10 mm (grey) + 6 mm clear	45	48	52	46	50	53	47	51	55	49	52	56	50	54	57	52	56	60	58	61	65	63	66	70
12 mm (grey) + 6 mm clear	48	51	55	49	53	56	50	54	58	52	55	59	53	57	60	56	59	63	61	65	68	67	70	74
10 mm (grey) + 10 mm clear	46	49	53	47	50	54	48	52	55	49	53	56	51	54	58	53	57	60	58	62	65	63	67	70
12 mm (grey) + 12 mm clear	48	52	55	50	53	57	51	55	58	53	56	60	54	57	61	57	60	64	62	66	69	68	71	75
6 mm (bronze) + 6 mm clear	37	40	44	38	41	45	39	42	46	40	43	47	41	44	48	43	46	50	46	50	54	50	54	57
6 mm (bronze) + 6 mm (bronze)	41	45	48	42	46	49	43	47	50	44	48	51	46	49	53	48	51	55	52	56	59	57	60	64
10 mm (bronze) + 6 mm clear	45	48	52	46	50	53	47	51	55	49	52	56	50	54	57	52	56	60	58	61	65	63	66	70
12 mm (bronze) + 6 mm clear	48	51	55	49	53	56	50	54	58	52	55	59	53	57	60	56	59	63	61	65	68	67	70	74
10 mm (bronze) + 10 mm clear	46	49	53	47	50	54	48	52	55	49	53	56	51	54	58	53	57	60	58	62	65	63	67	70
12 mm (bronze) + 12 mm clear	48	52	55	50	53	57	51	55	58	53	56	60	54	57	61	57	60	64	62	66	69	68	71	75
Reflective, laminated combinations																								
6 mm 17/24 (gold) + 6 mm clear	31	36	39	32	37	40	33	38	41	34	38	42	35	39	43	37	41	44	40	44	48	43	47	51
6 mm 16/22 (gold) + 6 mm clear	29	34	37	30	34	38	31	35	39	32	36	39	32	37	40	34	38	42	37	41	45	40	44	48
6 mm 39/34 (gold) + 6 mm clear	26	30	34	27	31	35	28	32	35	28	32	36	29	33	37	30	34	38	33	37	41	35	40	43
Reflective sealed units																								
47/33 (azure)	37						39	43								43	47							
36/25 (gold)	37						38	42								42	46							
33/25 (bronze)	36						38	42								41	45							
26/22 (gold)	44						46	50								51	55							

Reduce values by 2K if airspace is ventilated

Table III. Basic temperature differences—INNER GLASS of double glazing.
The shaded areas indicate, without further calculation, that the glass is safe from thermal fracture.

	Solar radiation intensity (W/m²) 750			775			800			825			850			900			1000			1100		
Diurnal temperature range (K)	10	15	20	10	15	20	10	15	20	10	15	20	10	15	20	10	15	20	10	15	20	10	15	20
6 mm clear float + 6 mm clear	13	15	17	13	15	17	13	16	18	14	16	18	14	16	18	15	17	19	16	18	20	17	19	21
Surface-modified float combinations																								
6 mm (bronze) + 6 mm clear	17	20	21	18	20	22	18	21	22	19	21	23	19	21	23	20	22	24	22	24	26	24	26	28
6 mm (bronze) + 6 mm (bronze)	25	27	29	25	27	29	26	28	30	27	29	31	27	30	31	29	31	33	32	34	36	34	37	38
10 or 12 mm (bronze) + 6 mm clear	19	22	23	20	22	24	20	23	24	21	23	25	21	24	25	23	25	27	25	26	28	27	28	29
10 mm (bronze) + 10 mm clear	21	23	25	22	24	26	22	25	26	23	25	27	23	26	27	25	27	29	27	29	31	29	32	33
12 mm (bronze) + 12 mm clear	22	24	26	23	25	27	24	26	28	24	26	28	25	27	29	26	28	30	29	31	33	31	33	35
Body-tinted float combinations																								
6 mm (green) + 6 mm clear	20	23	24	21	23	25	22	24	26	22	24	26	23	25	27	24	26	29	26	28	30	28	31	32
6 mm (green) + 6 mm (green)	30	32	34	31	33	35	32	34	36	33	35	37	34	36	38	36	38	40	39	41	43	43	45	47
10 mm (green) + 6 mm clear	24	26	28	25	27	29	25	27	29	26	28	30	27	29	31	28	30	32	31	33	35	34	36	38
10 mm (green) + 10 mm clear	25	27	29	26	28	30	27	29	31	27	29	31	28	30	32	29	32	34	32	35	36	35	37	39
6 mm (grey) + 6 mm clear	21	23	25	22	24	26	23	24	26	24	25	27	24	26	27	25	28	29	27	29	31	30	31	33
6 mm (grey) + 6 mm (grey)	31	33	35	32	34	36	33	35	37	34	36	38	35	37	39	37	39	41	41	42	44	44	46	48
10 mm (grey) + 6 mm clear	25	27	28	25	27	29	26	28	30	27	29	31	28	29	31	29	31	33	32	34	35	35	36	38
12 mm (grey) + 6 mm clear	26	28	30	27	29	30	27	29	31	28	30	32	29	31	33	30	32	34	33	35	37	36	38	40
10 mm (grey) + 10 mm clear	26	28	30	27	28	30	27	29	31	28	30	32	29	31	33	30	32	34	33	35	37	36	38	40
12 mm (grey) + 12 mm clear	27	29	31	28	30	32	29	31	33	30	32	33	30	32	34	32	34	36	35	37	39	38	40	42
6 mm (bronze) + 6 mm clear	21	23	25	22	24	26	23	24	26	24	25	27	24	26	27	25	28	29	27	29	31	30	31	33
6 mm (bronze) + 6 mm (bronze)	31	33	35	32	34	36	33	35	37	34	36	38	35	37	39	37	39	41	41	42	44	44	46	48
10 mm (bronze) + 6 mm clear	25	27	28	25	27	29	26	28	30	27	29	31	28	29	31	29	31	33	32	34	35	35	36	38
12 mm (bronze) + 6 mm clear	26	28	30	27	29	30	27	29	31	28	30	32	29	31	33	30	32	34	33	35	37	36	38	40
10 mm (bronze) + 10 mm clear	26	28	30	27	28	30	27	29	31	28	30	32	29	31	33	30	32	34	33	35	37	36	38	40
12 mm (bronze) + 12 mm clear	27	29	31	28	30	32	29	31	33	30	32	33	30	32	34	32	34	36	35	37	39	38	40	42
Reflective laminated combinations																								
6 mm 17/24 (gold) + 6 mm clear	16	18	20	16	18	20	17	19	21	17	19	21	17	20	22	20	22	24	21	24	25	23	25	27
6 mm 16/22 (gold) + 6 mm clear	15	17	19	15	18	19	16	18	20	16	18	20	17	19	21	18	20	22	20	22	24	22	24	26
6 mm 39/34 (gold) + 6 mm clear	17	19	21	17	20	21	18	20	22	18	20	22	19	21	23	17	20	21	19	21	23	21	23	25
Reflective sealed units																								
47/33 (azure)	14						15	16								16	18							
36/25 (gold)	13						13	15								15	16							
33/25 (bronze)	13						13	14								15	16							
26/22 (gold)	14						15	16								17	18							

Reduce values by 2K if the airspace is ventilated.

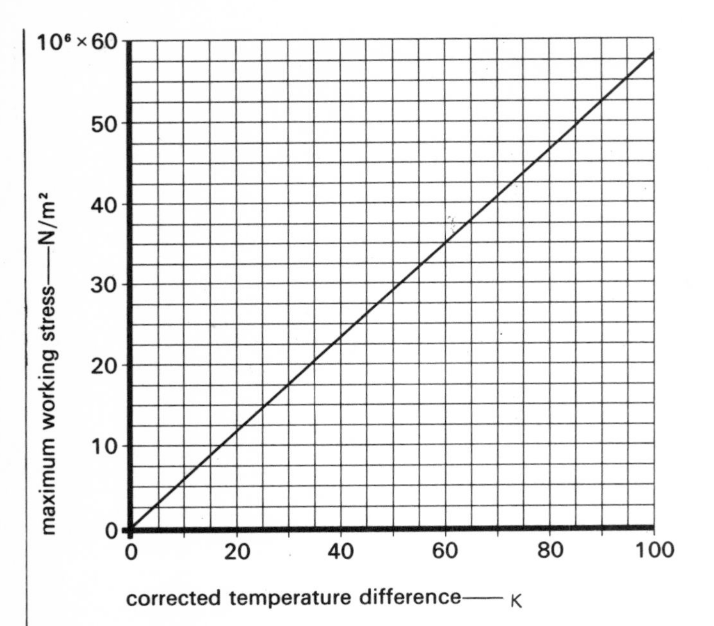

8 *Variation of maximum working stress with corrected temperature difference.*

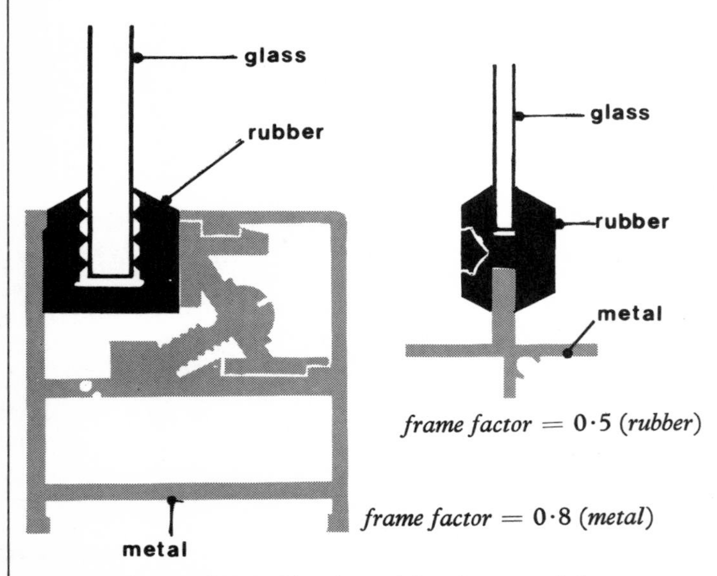

9 *Frame factors for combinations of framing materials.*

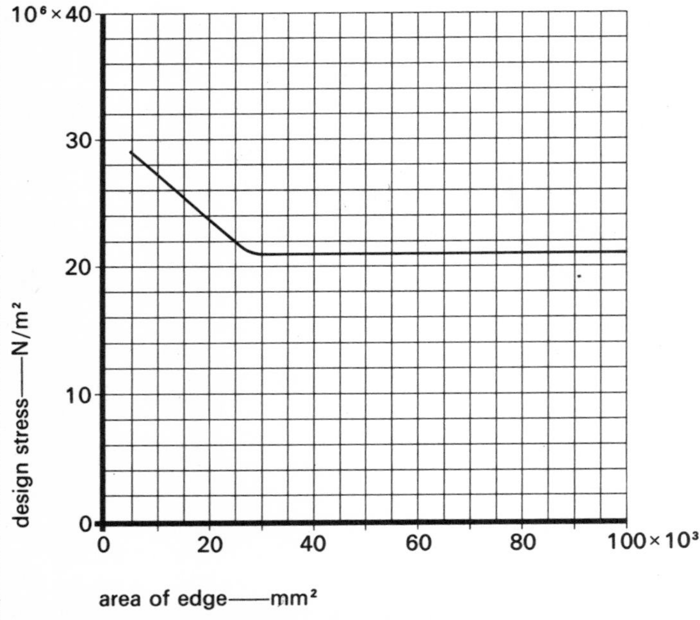

10 *Variation of design stress with area of edge.*

Recommendations

When the thermal safety of the glass in a particular window design has been assessed, if the glass is found to be at risk, all the factors shown in **11** should be studied to see which can be changed to reduce the working stress or increase the design stress. Some factors have more influence than others but may be fixed by decisions in other aspects of the overall design—appearance or environmental control, for example—in which case improvement will depend upon adjustment of the minor factors. Changing the size of the individual squares of glass or tilting the glass may have a small effect but the major improvements are likely to be found in changes to the design of blinds and frames.

Table IV shows the influence of various blind design factors, the smaller corrections indicating reduced risk of thermal breakage. It is generally better to seek improvement by ventilating the glass-blind space; changing the colour of a blind from light to dark usually conflicts with the needs for a highly reflective blind for effective control of solar radiation.

The effect of the material of the frame on the working stress is shown in Table V.

Heat loss from the room through metal window frames and condensation on the inner surfaces of the frames are often reduced by building a poorly conducting layer into the frame, as at A in **12**. Any such thermal resistance must extend to the outside of the glazing line so that the glass is effectively glazed in the warmer, inner members of the frame and not in the colder, outer parts.

The inner pane of sealed, double glazing units may be at risk in cold climates. The low thermal resistance of the edge seal tends to keep the edge of the inner glass at the outdoor air temperature while the central areas are exposed to the warm room air. The risk is reduced by any measure that encourages warm edges to the unit. The frames shown in **12** and **13** are effective. The central areas of the inner glass should be kept as cool as possible, so heating outlets should be directed away from the glass and there should be free ventilation of the space between the glass and any blinds or curtains. There should be no structural 'pockets', like that shown at the top of the window in **14**, where warm air can collect and raise the temperature of the glass.

Some of the solar control glasses can be toughened and this gives a means of raising the design stress and ensuring safety from thermal fracture. The appearance of the glasses is not affected by the toughening process so it is only necessary to toughen the glass for those façades that are at risk. It may also be advisable to toughen any squares of solar control glass that are drilled or notched because of the difficulty, sometimes

Table IV. Corrections to be added to the basic temperature difference if blinds are fitted. The term 'blind' includes all similar installations, such as curtains and venetian blinds with the slats full closed.

	Light blind K	Dark blind K
Single glazing		
Glass—blind space ventilated	5	3
Glass—blind space not ventilated	11	9
Venetian blind with slats at 45°	5	3
Double glazing, outer glass		
Glass—blind space ventilated	4	0
Glass—blind space not ventilated	4	0
Venetian blind with slats at 45°	4	0
Blind between glasses	12	9
Double glazing, inner glass		
Glass—blind space ventilated	6	4
Glass—blind space not ventilated	8	5
Venetian blind with slats at 45°	6	4

Table V. Frame factors

Framing material	Factor
Concrete	1·0
Wood	0·9
Metal	0·8
Plastics or rubber	0·5

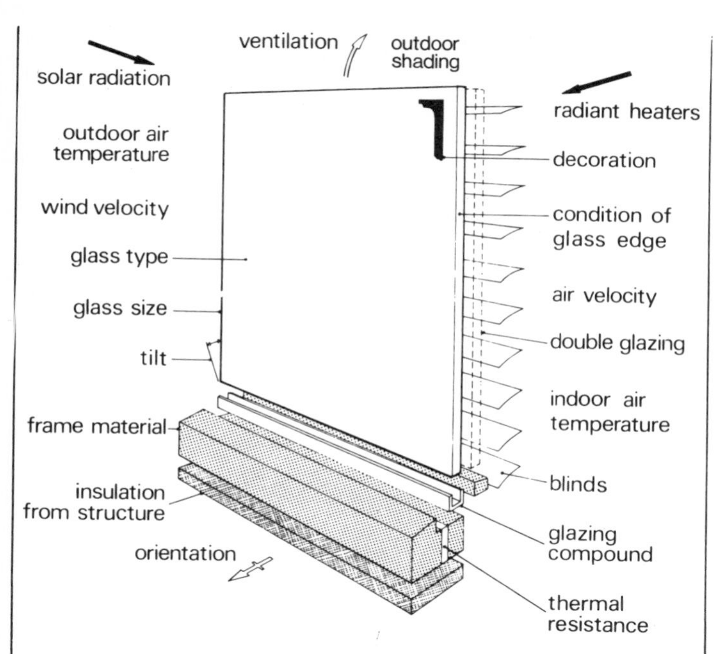

11 *The factors affecting thermal safety.*

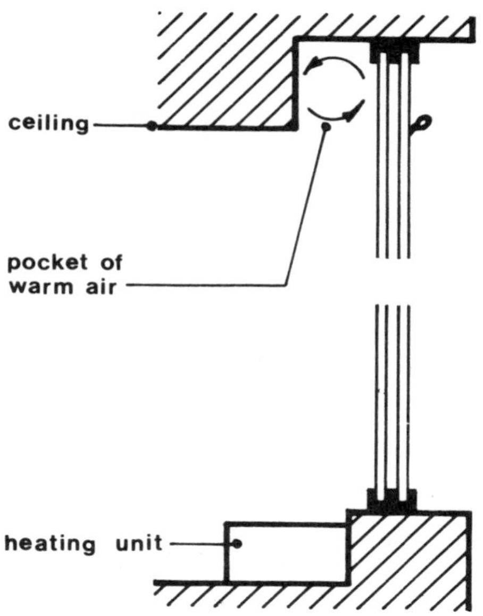

14 *Structural details that create pockets of warm air close to the glass should be avoided.*

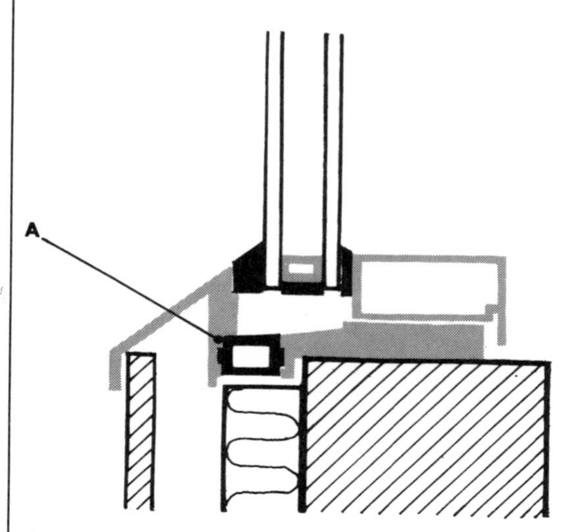

12 *Metal frame with thermal resistance.*

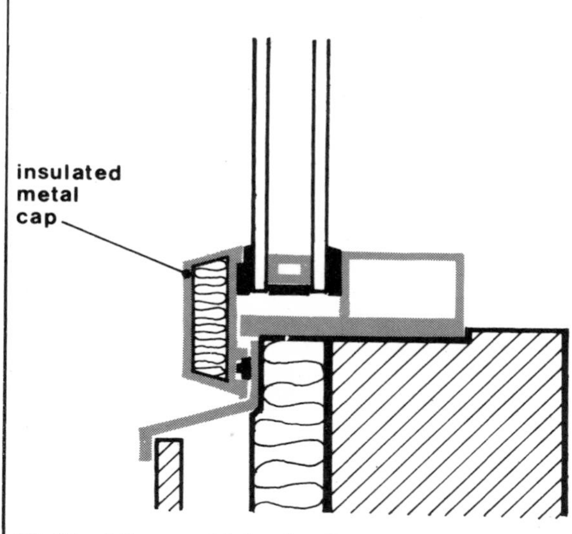

13 *Metal frame with insulated cap.*

experienced, of producing an edge of acceptable quality within the hole.

It is important to remember that the safety assessment is based on the behaviour of glass in good condition and properly glazed. Even if the glass is shown to be thermally safe, that safety depends on close adherence to the glazing procedures recommended. All necessary precautions must be taken to see that only glass with edges of an acceptable condition is glazed: the glass must be stored and handled so that no contact with hard bodies can damage the edges and each square should be carefully examined immediately before glazing so that any damaged glass can be rejected. When the edges of sealed double glazing units are protected by a metal tape the condition of the edges of the glasses cannot be seen although damage to the tape may indicate that the glass below has been damaged. Double glazing units should, therefore, receive special care in handling.

It is advisable to avoid the use of opaque or dark coloured labels, paint or decoration or any other detail that, by increasing the absorptivity of the panel, tends to create localised areas of high temperature and affect the stress distribution. Modifications of this sort effectively convert the treated areas of clear glass to heat absorbing glass and introduce the need for all the precautions discussed in this chapter.

Where solar control glasses are to be used in sliding panels or in sash windows there is always the possibility that, when windows are opened during sunny periods, the overlapping glasses will react as double glazing with little ventilation in the air space and this is the condition that should be assumed in assessing the thermal safety of the glass.

Some aspects of thermal safety—the use of double glazing units in sash windows, annealed glass used as cladding with a back-up wall behind, solar control glass used in roof glazing or other non-vertical situations, glass shaded by overhangs of more than 150 mm or by mullions that protrude more than 450 mm, glass with opaque labels or applied decoration—need a more detailed analysis than could be covered here. In these cases, or in any case where there is doubt or the prediction technique described gives a borderline result or a marginal rejection, the manufacturer should be consulted.

Glazing solar-control glass

The glazing procedure for solar control glasses should follow the recommendations of the British Standard Code of Practice CP 152, the Glazing Manual of the Glass and Glazing Federation (formerly of the Flat Glass Association), and the manufacturer's literature. The main points are outlined here.

Single glazing

Body-tinted glass

Edge clearance. Because all types of solar control glass absorb more heat than clear glass of the same thickness, they become hotter and expand more, so greater clearance at the edge is needed. The clearance should be at least 5 mm all round unless no dimension of the glass exceeds 750 mm when it may be reduced to 3 mm.

Front and back clearance. Non-setting glazing compounds can deteriorate if used in layers that are too thin. To ensure that the properties of such compounds are retained minimum clearances of 3 mm should be allowed between the glass and the upstand of the rebate and between the glass and a glazing bead.

Edge cover. The edge cover should be sufficient to retain the glass safely in the frame. Thermal safety is calculated for the most stringent edge cover and experiment has shown that, within limits that are consistent with safe retention of the glass, the improvement due to variation of the width of the edge cover is negligible.

Glazing beads. The section of a glazing bead should be such as to give a height equal to that of the back check of the frame and a width adequate to resist displacement under wind loading. The securing points should be positioned to prevent warping of the beads.

Setting blocks. Setting blocks are the means by which the glass is centralised in its frame and the edge clearances are maintained. They should be of a non-absorbent, resilient material. Two setting blocks on the bottom frame member are normally required for each square of glass (see Chapter 1, Function 9, **2**). They are not necessarily of equal thickness; the thicknesses should be adjusted to make a uniform edge clearance all round the glass. The length of the blocks may be from 25 to 150 mm according to the size of the glass: the width is chosen to suit the glazing system but is never less than the thickness of the glass.

Distance pieces. Distance pieces ensure that adequate front and back clearances are maintained and that glazing compounds are not displaced. They should be of a non-absorbent, resilient material and positioned opposite to each other at bead fixing points.

Location blocks. Location blocks are used in opening lights at the vertical and top edges to keep the glass centred in the frame. They should be of non-absorbent, resilient material and are usually about 25 mm long except in reversible, horizontal pivot windows where they should be of similar size to the setting blocks.

Non-setting compounds. Non-setting glazing compounds (hand-grade, gun-grade or strip materials) allow relative movement between the glass and the frame. Setting blocks and distance pieces must be used, but with some strip materials that are load-bearing the distance pieces may be omitted. Preformed, extruded, mastic strips fill much of the back and front clearance spaces but should be top-pointed.

Sealers. All absorbent frames and beads should be treated with a sealer to prevent deterioration of non-setting compounds. Sealers compatible with the compounds are available from compound manufacturers. Primers are not sealers.

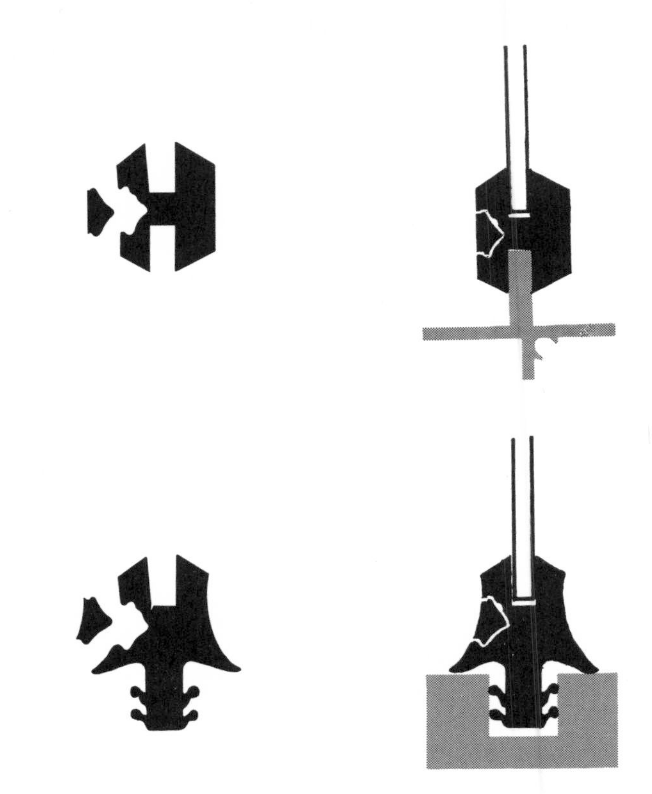

15 *Structural gaskets.*

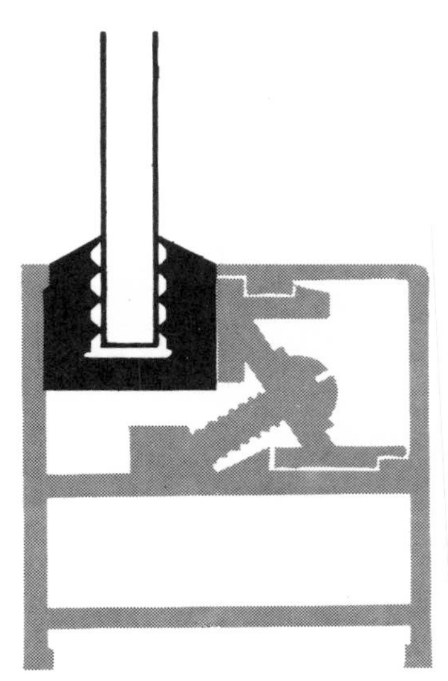

16 *Strip gaskets.*

Gaskets. For the gasket glazing of solar control glasses only the material specified in British Standard BS 4255 should be used. Of these, the most readily available is neoprene.

Structural gaskets. The glass is retained in a structural gasket by inserting a 'zipper' strip into one side of the gasket to compress the rebate flanges onto the glass, **15**.

U-section channel gaskets. U-section channel gaskets are used in conjunction with compression bead systems for site installations and with the 'knock-on' system for factory glazed frames.

Strip gaskets. Strip gaskets are usually rectangular or wedged in section and the seal is achieved by the use of compression beads as shown in **16**.

Glass edges. With two minor exceptions, a clean wheel-cut edge is the most satisfactory for all solar control glasses in all glazing systems. It is the strongest edge that can be achieved in practice and so gives the least risk of thermal fracture.
When the thicker heat absorbing glasses are cut it is not always possible to avoid the appearance of 'feather' on the cut edges. A small amount of feather is acceptable but glass should not be glazed if the edges carry extensive feather.
Solar control glasses must not be nipped to size and any squares with 'vented' edges must not be accepted for glazing.
The first of the exceptions mentioned above is that laminated glasses may be supplied with worked edges: the second that glazing gaskets may be damaged by the sharp edges of a clean cut. Where clean cut edges are not permissible, arrises should be created by a wet process, working parallel to the edge and not across the thickness.

Handling. Care must be taken to protect the edges of glass from damage. Glass should never be allowed to stand on or lean against materials that may damage the edges or surfaces.

Surface modified glass
Surface modified glasses should be glazed with the modified surface to the interior of the building. In all other respects the glazing recommendations for body-tinted glass should be followed.

Laminated glass
Laminated glasses should be glazed in accordance with the recommendations for body-tinted glass, but the glazing compounds chosen must be compatible with the laminating interlayer of the glass. In any glazing system the edge of the glass must be protected from contact with water.

Multiple glazing
The glazing techniques for multiple glazing units incorporating solar control glasses are generally the same as those for single glazing but three points deserve special attention.

Gasket glazing. Gasket glazing systems may be used with multiple glazing units provided that the design of the gasket and the frame prevents water from lodging at the edges of the unit. This normally requires the unit to be supported on setting blocks and the gasket to be provided with drainage.

Edges of units. Before a multiple glazing unit with protective metal tape on the edges is accepted for glazing the metal tape should be examined. Any damage to the tape may indicate underlying damage to the edge of the glass.

Setting blocks. With multiple glazing units, setting blocks should not be less than 40 mm long.

Appendix I

Suspended glass assemblies

A matrix of toughened glass plates sealed together and hung from the building structure to form a complete glass façade. This is the basic design concept of suspended glass assemblies. They allow the architect or designer to glaze large openings in buildings without the use of frames or mullions of any kind, offering the ability to create space and light in buildings with a minimum of visual barriers.

Essentially, the systems comprise a series of specially processed and toughened glass plates bolted together at their corners by means of small metal 'patch' fittings. Plate to plate joints are sealed with a silicone building sealant and toughened glass stabilisers are used at each vertical joint to provide lateral stiffness against wind loading. The assembly so produced is suspended from the building structure by hangers bolted to its top edge, and is sealed to the building in peripheral channels by means of neoprene strips or non-setting compound. **1** shows a typical assembly design.

The use of toughened glass and a suspension system gives a range of important benefits to assemblies which can be categorised under three main headings: *versatility*, *security* and *safety*.

Versatility

Because drilled holes are used to hang the assembly from the building structure, spectacular façade heights can be achieved. Single assemblies can be designed up to 20 m in height on a 1·5 m module, and up to 23 m on a 1·2 m module. Assuming an adequate main building structure, any height can be specified using multiple assembly designs. And, of course, any length is possible.

For example, one single vast assembly 305 m long and 16 m high forms the complete façade of a new office complex at Ipswich in England, and two tiers of spectators will watch racing in comfort behind a multiple assembly 134 m long and 20 m high at the new race grandstand at Louisiana Downs, USA.

With the right glass thickness and proper stabiliser configurations, assemblies can be designed to satisfy virtually any performance requirement. For example, the assembly for the Centre Point Building in London was designed to withstand wind pressures of 3200 N/m^2 and was tested to over 3800 N/m^2. Through extensive research and engineering work, much of it in collaboration with Birmingham University in England, a sophisticated theoretical understanding of suspended assembly design has been developed. A suite of specially written computer programs can predict, with a high degree of accuracy, the stress patterns and deformations which would occur within any system element. In combination with unrivalled practical experience, this gives the system a very high degree of flexibility, enabling it to cope with virtually any design criterion. A good example of this was the Standard Bank Building in Johannesburg. Because of the very large movements resulting

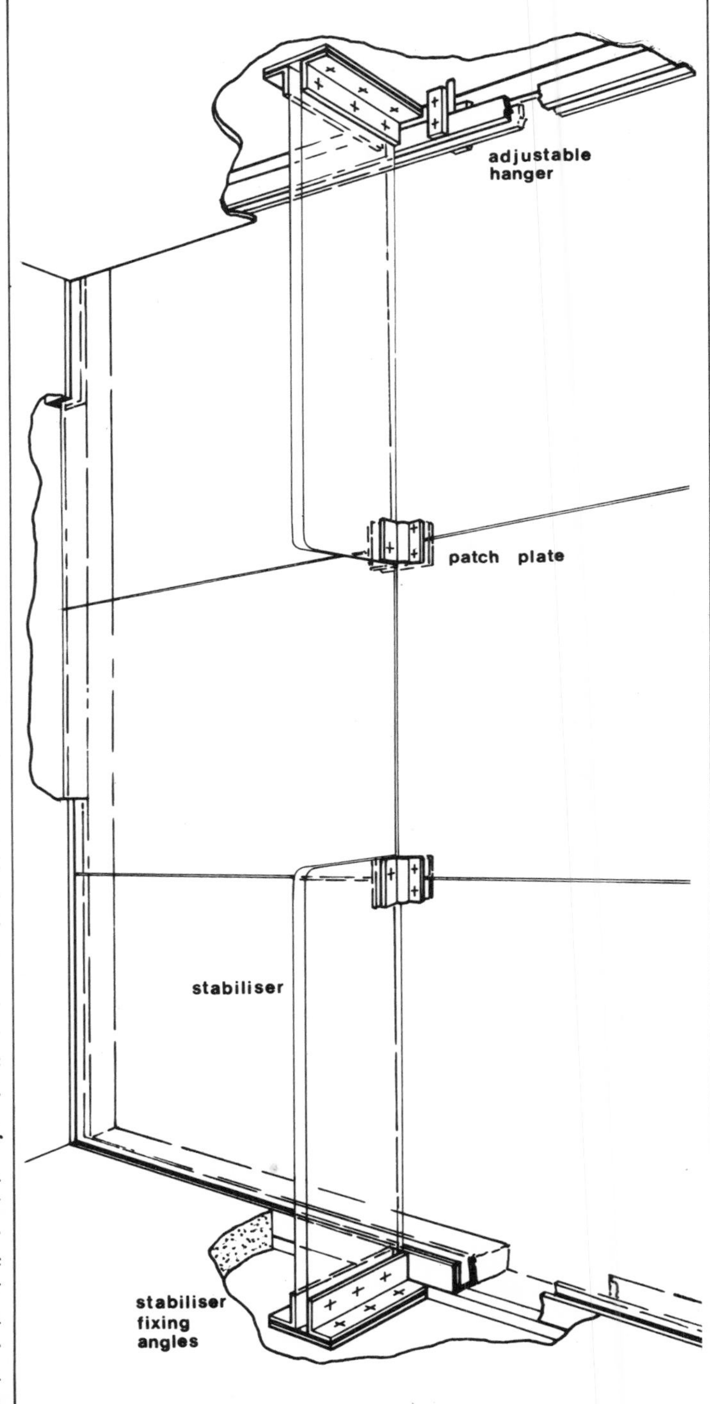

1 *Typical assembly design.*

from its unusual structure, a completely novel spring suspension system was produced.

Toughened solar control glasses can be used where required and access doors, air conditioning vents, etc, can be built into any assembly design.

The design of the system is such that it can be easily and speedily erected using conventional handling equipment.

Security

Because toughened glass is used, drilled holes can be employed to permit a simple and reliable method of attachment to the building structure and to other plates.

Toughened glass has a very high thermal strength. This is often required since large surface temperature differentials produced in severe climatic conditions can result in high stressing. This is particularly important when environmental considerations dictate the use of solar control glasses.

Toughened glass can withstand comparatively high localised stress concentrations, and therefore, extremely small metal 'patch' fittings can be used employing drilled holes. With annealed glass, the use of drilled holes is not possible, and pure frictional or adhesive concepts of design are necessary.

The concept of design ensures that the façade is at all times floating in peripheral channelling and problems which might arise due to differential movement between component parts are eliminated. Assemblies can, therefore, be used to full advantage when vibratory or seismic forces must be taken into account in the design.

During 1974 the independent, Government-funded Agrément Board in the United Kingdom, carried out a series of exhaustive tests on a full scale 9·1 m × 9·1 m assembly design. The test reports produced gave complete verification of the design technology.

Safety

In the unlikely event of stabilisers or plates being broken, the assembly will stay intact. It cannot collapse nor can individual plates move. In systems where the plates are held in position by friction only, it is possible for a plate to slip.

If broken, toughened glass shatters into a mass of small particles. These are not dangerous because they are small and because they do not have the sharp edges characteristic of annealed glass fracture.

Should any plate or stabiliser require replacement this can be achieved with minimum inconvenience.

Principles of design

More conventional forms of structural work are built from the ground upwards. Glass assemblies, however, are normally made from 12 mm thick toughened glass which, in comparison with most other structural materials, is lightweight and flexible. This means that for tall constructions, to build from the ground floor upwards could result in buckling of the lower plates under the weight of the glass above.

To overcome this problem, assemblies are constructed from the top downwards. This is the main concept of design. The assembly is suspended from the building structure and at all times the façade is allowed to 'float' in the peripheral channelling used for lateral restraint and weather sealing.

This section examines the principles involved in the major areas of design and illustrates details of the major system elements.

Suspension of the glass

It is not possible at the present time, for the building industry to provide structures that have the same dimensional accuracy as the glass used in suspended assemblies. The main support beams may not be true longitudinally and they can sag under their own weight. For concrete structures the effects of short and long term creep can be substantial.

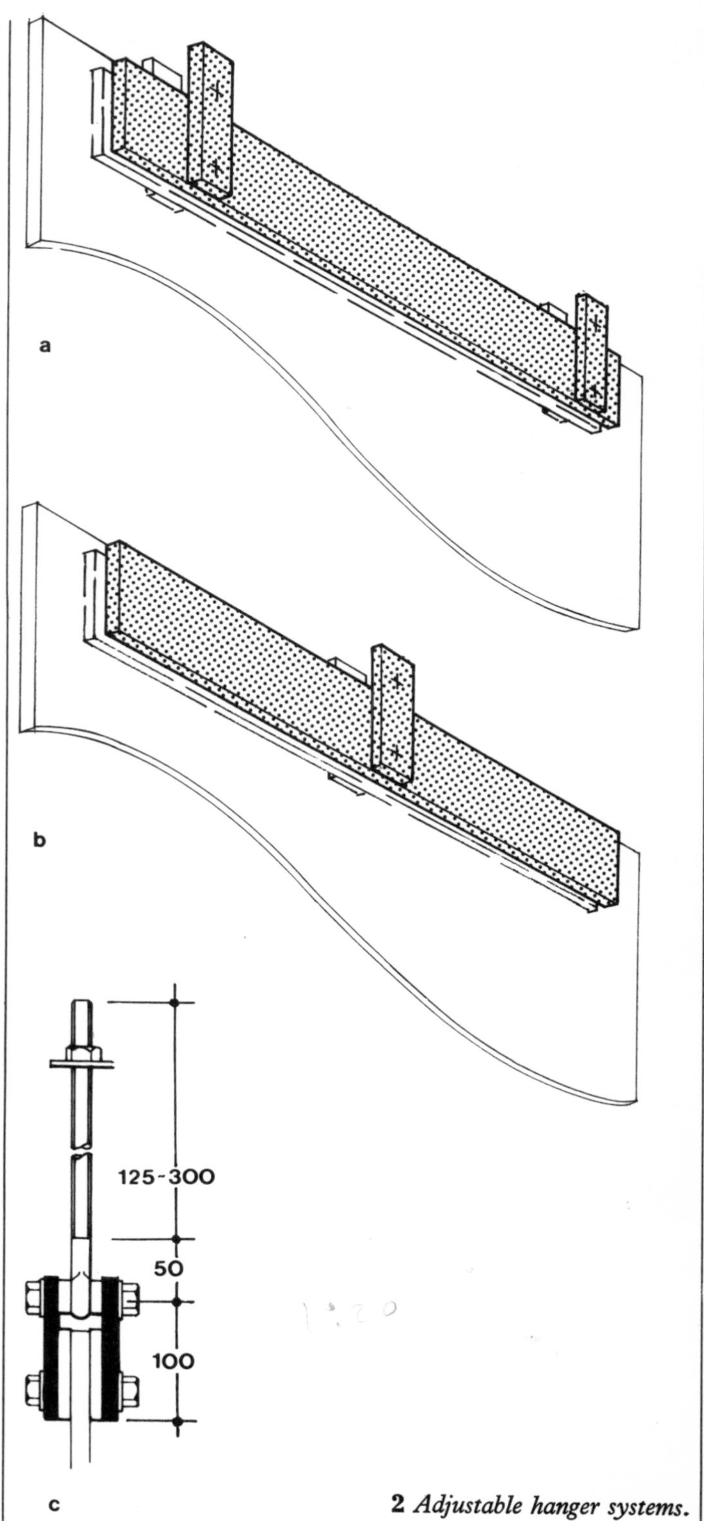

2 *Adjustable hanger systems.*

For these reasons, connection of the façade glass directly to the main structure is rarely a practical proposition since:

- Each vertical run of glass plates must be plumb.
- Adequate clearance between individual glass plates must be maintained.

It is necessary, therefore, to provide some form of adjustment to enable these criteria to be met. This is provided by the use of standardised adjustable hanger systems.

In the more usual form of hanger system shown in **2a** two suspension points are used per glass plate. One suspension point is used, however, where excessive deflections of the supporting structure occur. This ensures that the façade glass plates always remain in a vertical plane under large live load deflections of the support beams. This is illustrated in **2b**.

2c shows the critical dimensions of these adjustable hanger systems.

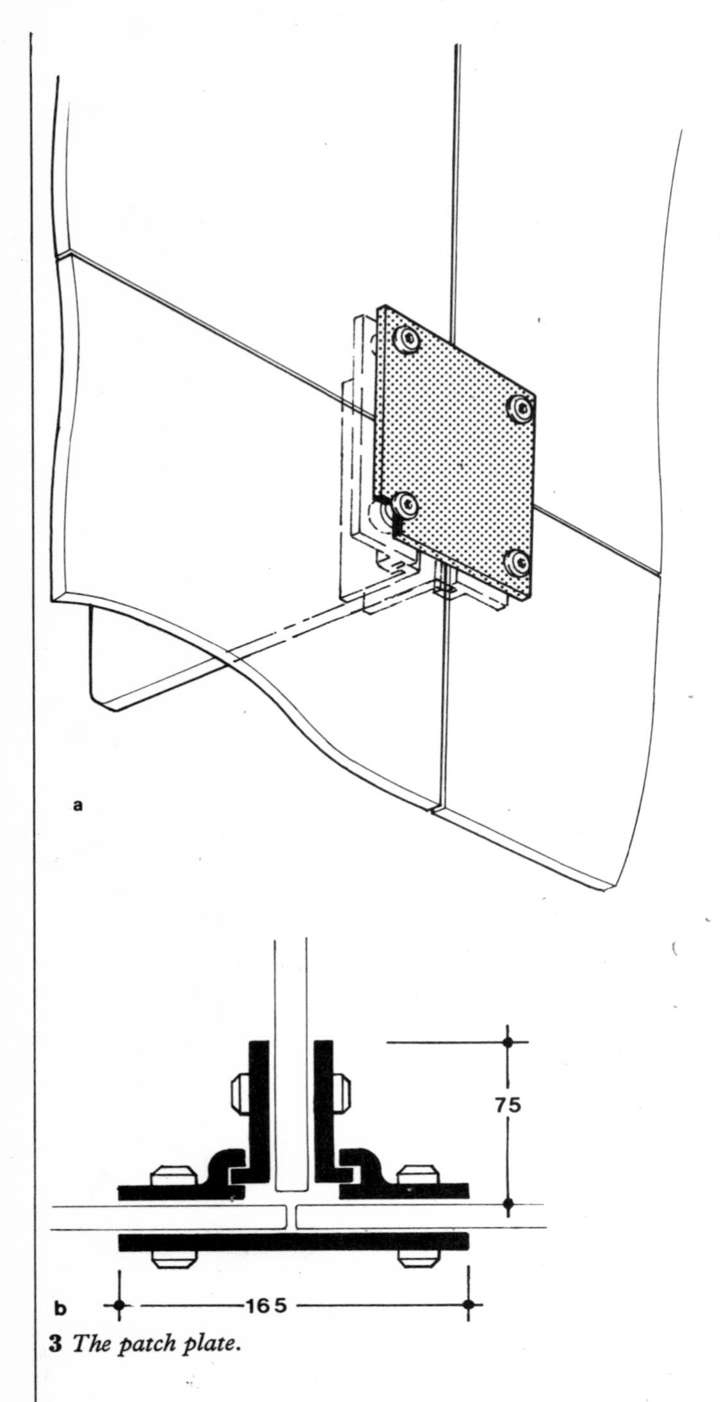

3 *The patch plate.*

By using suspension hangers, compensation can be provided for inaccuracies in either horizontal or vertical directions. The effects of long term creep can also be accommodated. Vertical adjustment is provided by simply turning a nut at the top of the suspension bolts. Horizontal adjustment is normally provided by means of oversize holes in the structure through which the suspension bolts pass.
Careful attention is paid to the environment to which the hanging system will be subjected and specially chosen protective finishes are applied. Each glass plate in the top tier of an assembly is suspended and will, therefore, be subjected to the dead weight of all those beneath.
Because of these dead weight loads, there is a limit to the height to which an assembly can be constructed. The height is dependent on the width of the individual glass plates chosen, which in turn, must be compatible with those calculated for resistance to wind loadings.

Method of securing glass plates together

A simple and reliable method of attachment at the corners of each glass plate is used. Adjacent plates are connected together by means of a metal 'patch plate' shown in **3a** and **3b**.
The design of the patch plate is an optimum between what is acceptable in structural terms and what is acceptable aesthetically.
The patch plates have been designed to ensure:

- Removal of maximum stress areas from around the holes in each glass plate.
- Moderation of the clamping forces necessary, thus alleviating high-peripheral stresses.

Design for acceptable glass plate size

The method used for determining the safe load capacity of various glass plate sizes, when subjected to wind load, has been obtained from both theoretical and experimental approaches. The load bearing capacity of each glass plate in the façade is dependent upon the way in which it is supported and hence on the configuration of the assembly.
The results of the design work are summarised in 'System design aid', below. Using the tables the maximum permissible size of glass plate which can be used with varying assembly configurations and wind loadings can be determined.
The 'Glass specification' section indicates the maximum sizes of glass plates that are available.

Design of the stabilisers

Stabilisers are used to provide lateral stiffness against wind loading and their various designs are shown in **4a** to **4e**.
4a and **4b** illustrate *cantilevered stabiliser systems*. Type **a** is normally used for assemblies with a maximum height of 9 m and type **b**, where full space utilisation at the floorline is wanted. The stabilisers can be tapered if required.
4c and **4d** illustrate *continuous stabiliser systems*. The 'propped cantilever' type **c** is used for tall constructions or where design wind loadings warrant its use. The 'pin jointed' or 'simply supported' type **d** is also used for tall constructions but has the advantage of eliminating torsional forces at the head of the structure. Instead, only lateral loads are exerted which makes the system ideal for lightweight constructions. The *balanced beam* type **e** can be used where a mezzanine floor is close enough to the façade to allow it. With this system, good floor space utilisation is possible and under some circumstances stabilisers can be eliminated altogether, rods being used to tie each 'patch plate' on the façade back to the floors.
The width of stabiliser depends upon the configuration chosen, and in design the following considerations are taken into account.

- The wind loading to which the assembly is designed and the size of individual glass plates. These will dictate the live loads acting on the stabilisers.
- The length of the stabiliser and hence its resistance to buckling.

The design of the stabiliser arrangement has been subjected to rigorous experimental work involving both load bearing and fatigue, or cyclic loading conditions. A summary of the results is shown in 'System design aid'. Using the tables, the width of the stabiliser required for particular configurations and design wind loads can be determined.
5 illustrates the fixing angle which ensures that the root of each cantilevered stabiliser is effectively clamped. The design, subject of a great deal of experimental work, prevents rotation of the stabiliser within the fixing angle by using sufficient clamping forces to maintain a high coefficient of friction at the glass/gasket/metal interfaces. In this way, bearing and shearing forces are removed which would otherwise act at the hole positions. This is a guiding principle in any clamp design.
It is necessary for some configurations to incorporate spliced joints into the stabilisers since it is not possible to manufacture the continuous lengths that are sometimes required. **6a** and **6b** illustrate typical designs based on similar principles to those for the fixing angles. Again these designs have been extensively tested under both sustained and cyclic load conditions.
The stabiliser retaining box shown in **7** is necessary for

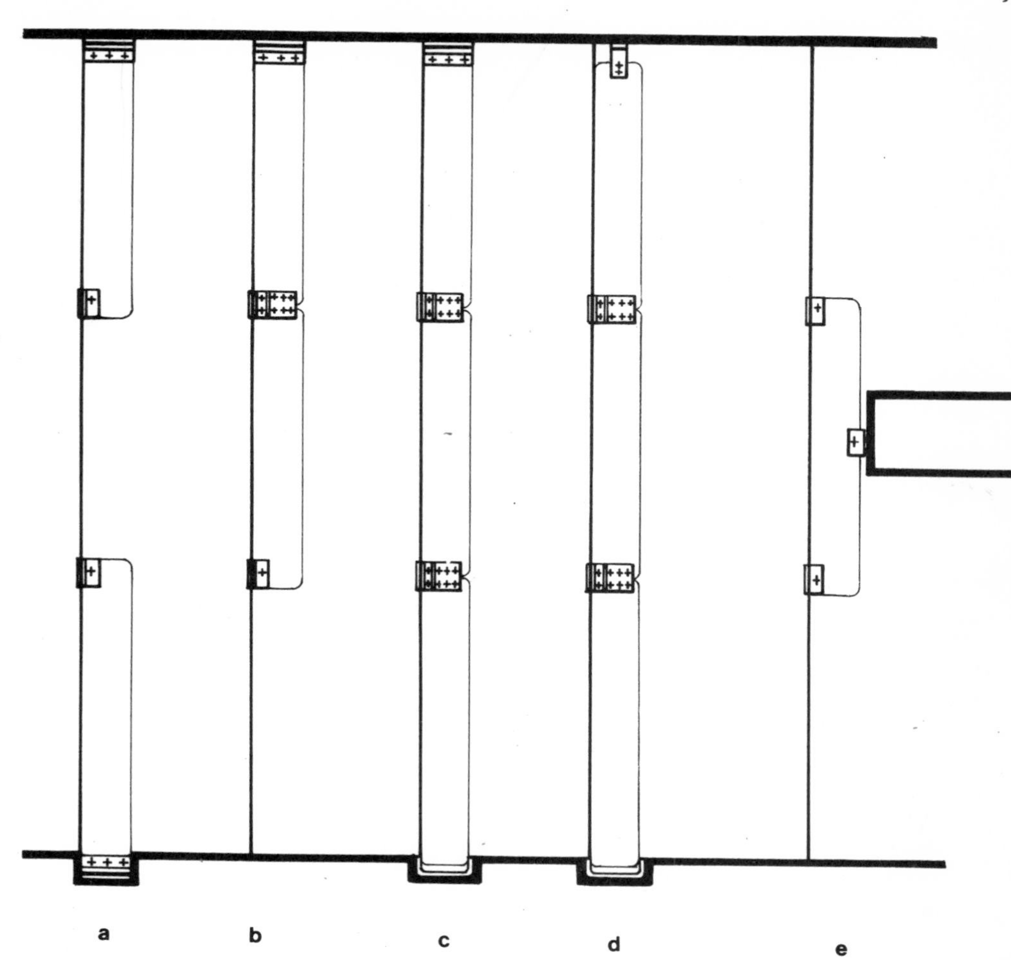

4 *Alternative stabiliser configurations.*

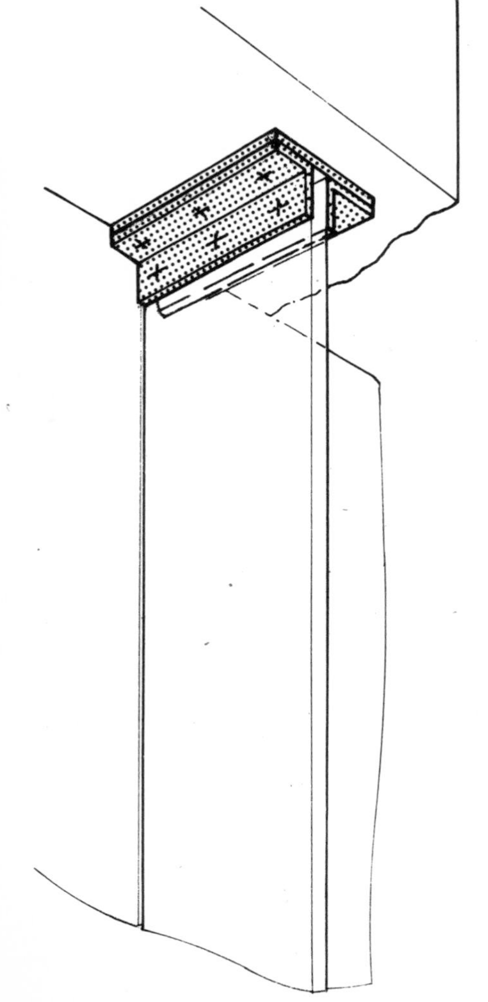

5 *Root fixing angle.*

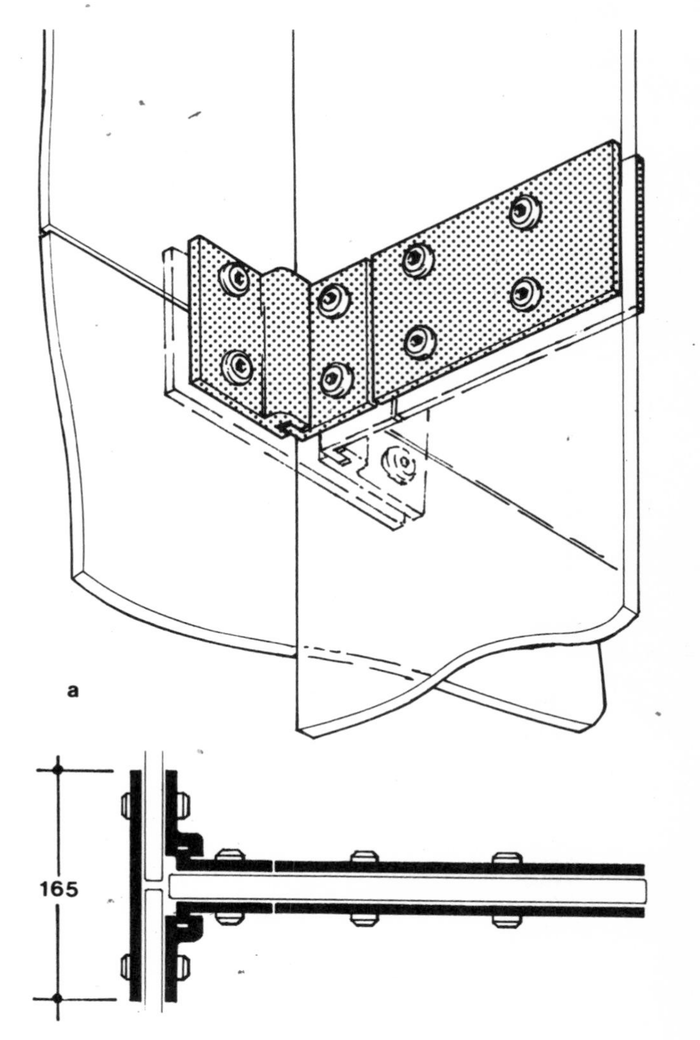

6 *Spliced joints.*

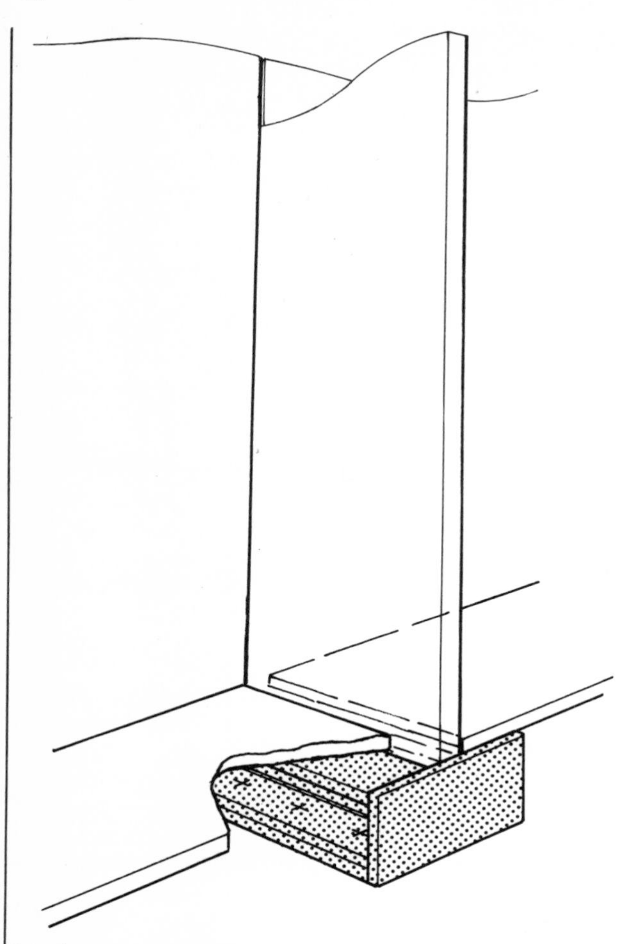

7 *Stabiliser retaining box.*

'propped cantilever' or 'pin-jointed' systems, to fix the stabiliser at the floorline while allowing in-plane rotation. The arrangement, however, prevents out-of-plane rotation.

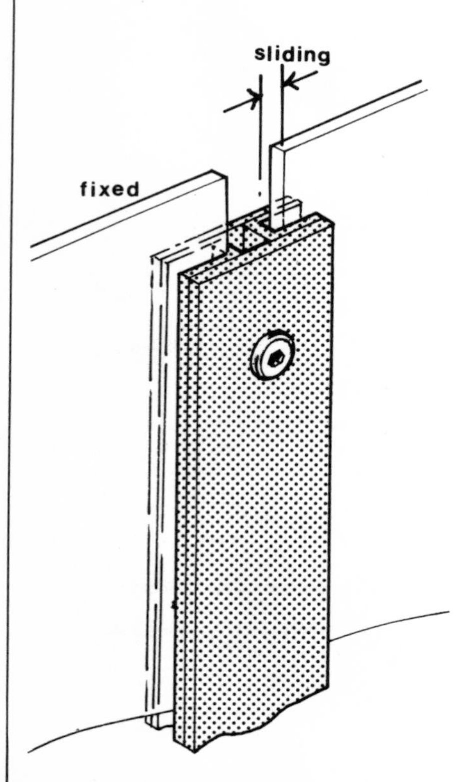

8 *Expansion joint.*

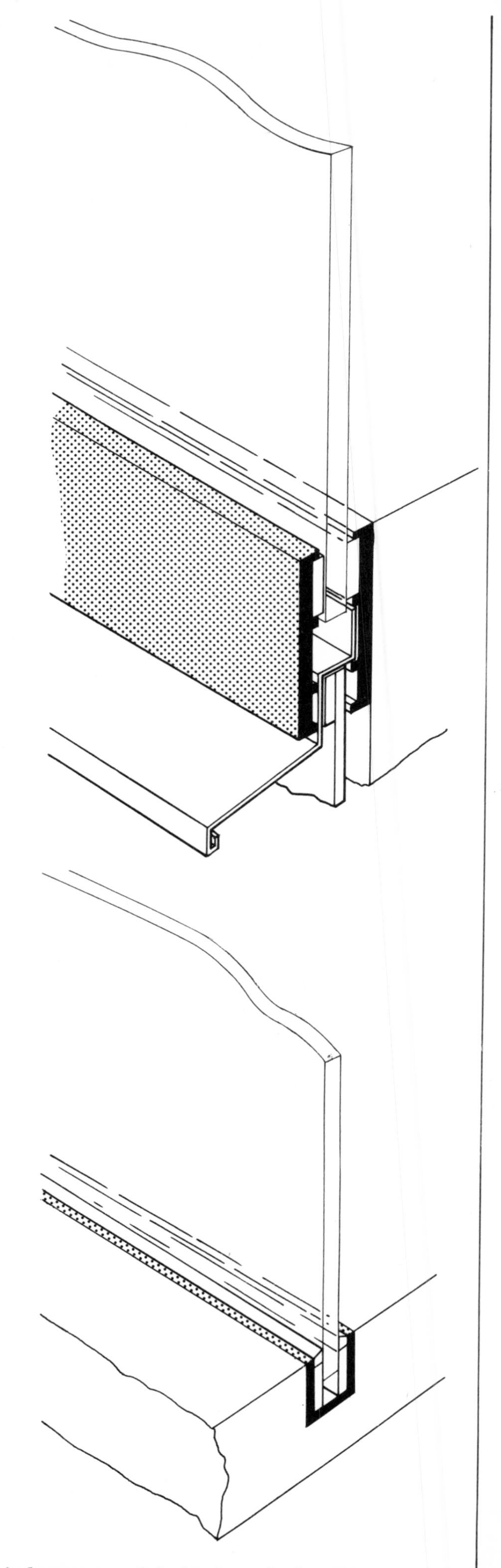

9 *Typical treatments of the façade at the floor line.*

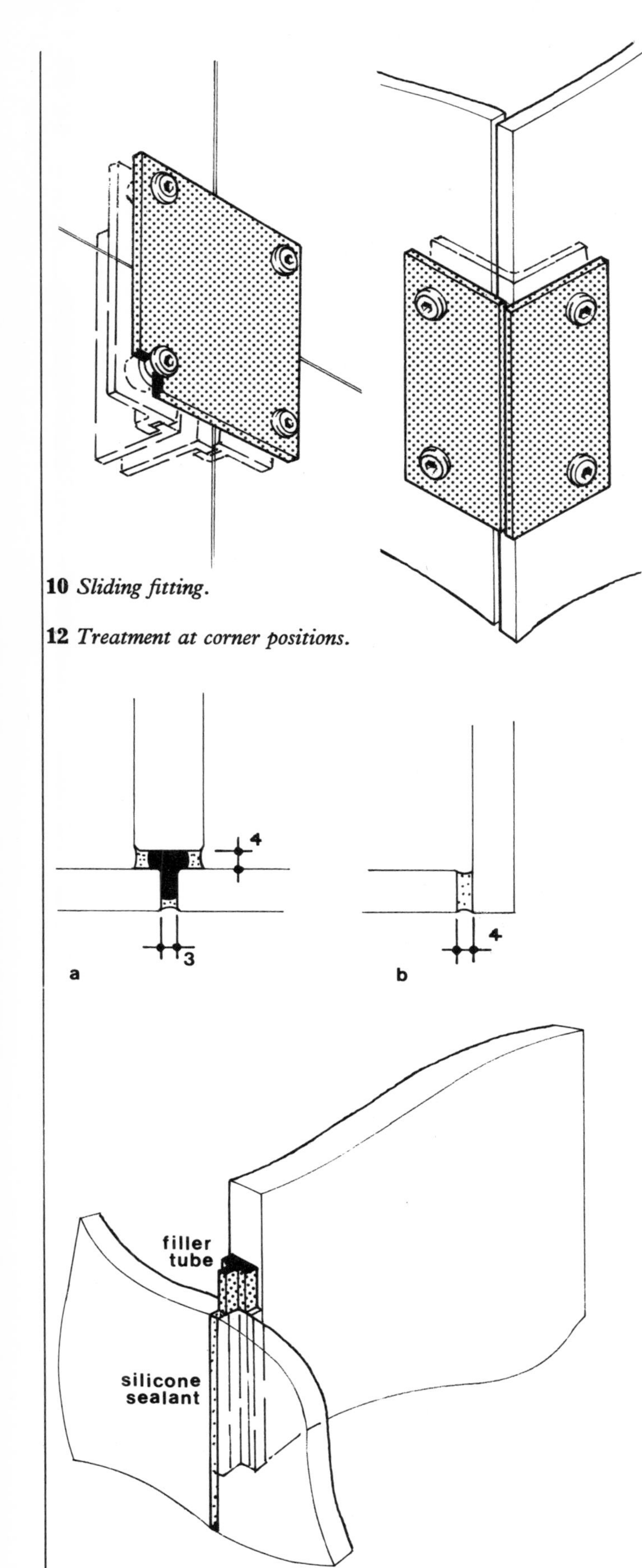

10 *Sliding fitting.*

12 *Treatment at corner positions.*

11 *Weather sealing joints.*

Thermal expansion and live load movements

Expansion joints can be designed into the system which will overcome any problems which may occur due to thermal movement. A typical example is shown in **8**, but many variations are possible. One side of the joint is flexibly fixed to the façade glass using an adhesive. The other side is not fixed and sufficient clearance is allowed for the glass to expand and contract within the channel without fouling. The amount of clearance will depend on the length of façade under consideration and the climatic conditions involved.

The joints would normally coincide with expansion joints in the structure itself, but they can be designed in almost any required position. The coefficient of expansion of glass is so small, however, that resultant thermal movement will be small, in comparison with other building materials.

The problem of differential thermal movement between the glass façade and the building structure, can be overcome by ensuring that the peripheral channelling is deep enough to accommodate any possible movement of the façade. In this way 'bottoming' of the glass edges is prevented and the façade always remains in a suspended state.

9a and **9b** indicate ways in which the façade can be treated at the floorline.

The method shown in **9a** is normally used in high rain exposure situations where there is an unacceptable risk of water penetration at the floorline.

The amount of clearance between the bottom edge of the façade and the bottom of the channel is dependent upon the amount of downward movement expected both due to live loads and thermal loads. The method of glazing is also dependent upon this movement. If it is small, then non-setting compounds for full bedding can be used. However, if the amount of movement is likely to be large then specially compounded neoprene strip dry glazing is used. Using neoprene gasketing, the façade is allowed to move in a positive or negative direction, and draining holes are provided to ensure that water build up in the channelling does not occur. Similar treatments can be used at the jambs.

Differential thermal movement between the glass plates and stabilisers will also occur. The façade will expand downwards whilst the stabilisers at the sill level, which can be fixed, will expand upwards. To compensate for these opposed forces, a sliding fitting is used at all façade/stabiliser connection points. Illustrated in **10**, these fittings also compensate for differential movements caused by live loading.

Weather sealing

Weather sealing is carried out at all joints in the façade using a gunned silicone building sealant. In the design calculations the structural properties of the sealant, in providing greater stiffness to the façade, are not recognised. However, from extensive laboratory and on site testing it is known that the sealant does improve the load bearing capabilities of the façade and its use is, therefore, an added safety factor in design.

11a and **11b** detail the types of weather sealing joints used in assembly designs. At the vertical joints, the void between the back of the façade glass plates and stabilisers, is filled using a polypropylene filler tube. This tube is used to facilitate gunning of the silicone sealant. The sealing treatment at the floorline has already been discussed with reference to **9a** and **9b**. **12** illustrates the type of fitting used at corner positions where return plates are involved in the design.

Design parameters

Prior to any detailed study for a particular application, there is a minimum quantity of information which is required:

- Details of opening to be glazed—normally drawings (plans, elevations, sections) are required.
- Wind loading to which the façade must be designed.
- Anticipated movement of the supporting structure due to:
  - live and dead loads
  - creep
  - thermal loadings
  - building tolerances.
- Environmental control requirement if any.
- Deflection restrictions, if any.
- Necessary compliance with local building codes.
- Any possible divergence from the use of standard fittings.

The information above would be required for a detailed design to be carried out. However, it is understood that much of this information may not be available at the initial enquiry stage and preliminary designs and cost estimates can be prepared as long as:

- opening size
- wind loadings
- aesthetic requirements are known.

It is the responsibility of the architect and/or structural engineers for the project to ensure that the building fabric is capable of supporting the façade.

System design aid

This section contains a set of tables which will allow the designer to determine the assembly configurations permissible for the project under consideration. Added to this, **13** enables the maximum glass plate widths that can be used for various assembly heights to be determined.

The tables, however, should only be regarded as a design *guide*. More comprehensive studies can only be undertaken by the manufacturer when the detailed information set out in 'Design parameters' is available. It should also be added that the figures given are only relevant to the performance of specially processed and toughened glass.

The following sequence should be followed when using the tables:

1. With a knowledge of the wind loads to which the structure must be designed and the stabiliser configuration preferred, **4**, the maximum glass plate areas and the widths of stabilisers required can be determined.
2. Knowing the total assembly height, **13** can be used to find the maximum width of each individual glass plate.
3. Consulting the section 'Glass specification' the maximum sizes available in the glass types under consideration can be found.

13 *Assembly height limitations for 2 point suspension system. Note: For the frictional grip suspension system the heights can be increased by the modular height up to a maximum of* 3·0 *metres.*

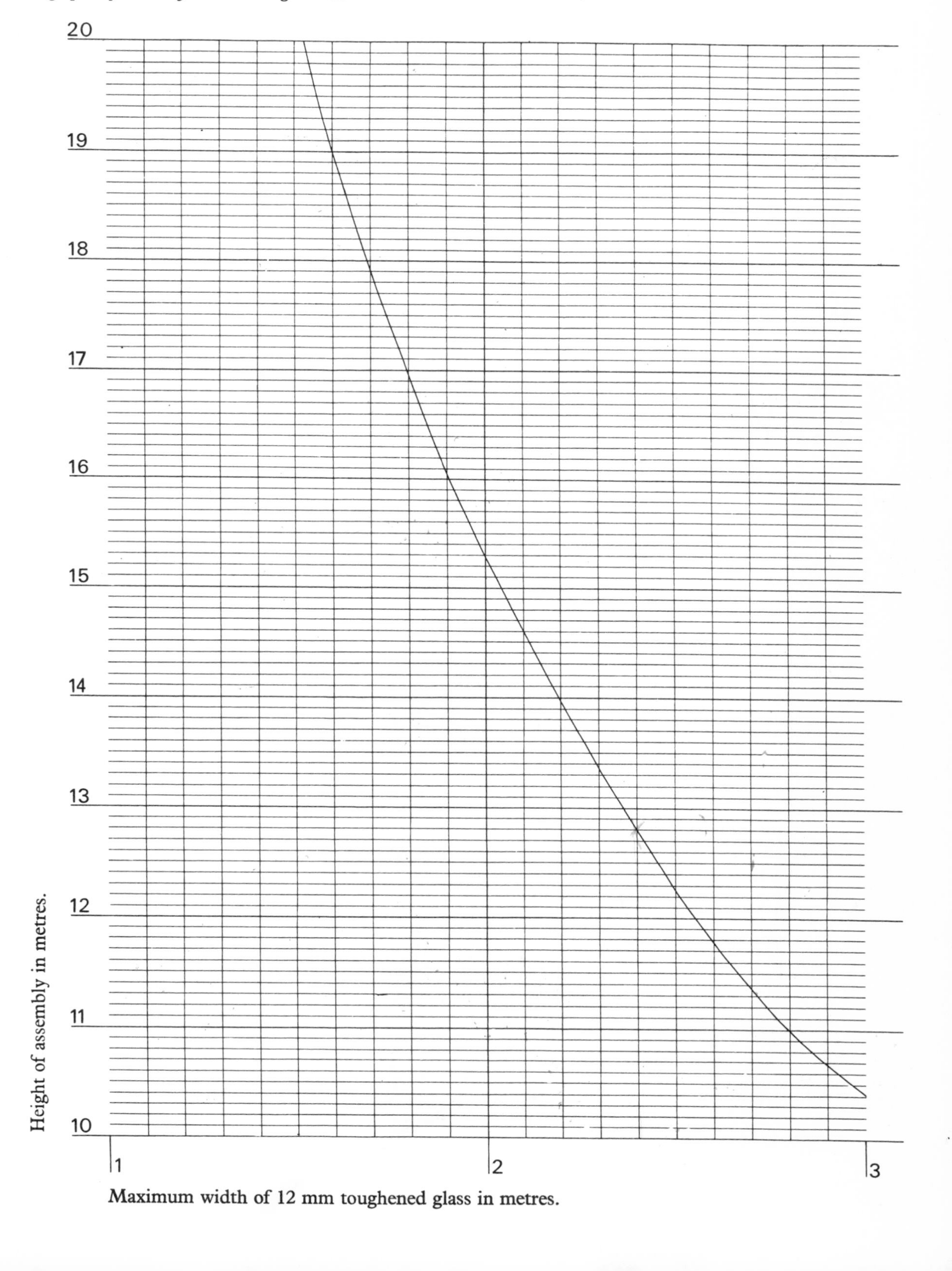

Cantilevered stabiliser system

Height of assembly m	Number of tiers	Area of individual plates m²		Stabiliser arrangement	Width of stabiliser required mm
Wind load 700N/m²					
5·0	2 min.	7·4 max.	0	Cantilevered from top length = 2·5 m	406
7·5	3 min.	7·4 max.	0	Cantilevered from top and bottom length = 2·5 m	406
7·5	3 min.	7·4 max. 4·6	4·6 0	Cantilevered from top length = 5·0 m	508 406
7·5	4	5·6 max.	0	Cantilevered from top length = 3·75 m Cantilevered from bottom length = 1·9 m	406
9·0	3 min.	7·4 max. 6·5	6·5 0	Cantilevered from top length = 6·0 m	610 508
9·0	3 min.	7·4 max.	0	Cantilevered from top and bottom length = 3·0 m	406
9·0	4	6·5 max. 5·6	5·6 0	Cantilevered from top length = 6·75 m	711 610
9·0	4	7·0 max. 4·6	4·6 0	Cantilevered from top length = 4·5 m Cantilevered from bottom length = 2·25 m	508 406
12·0	4 min.	7·4 max. 5·6	5·6 0	Cantilevered from top length = 6·0 m Cantilevered from bottom length = 3·0 m	610 508
12·0	5	6·0 max. 4·6	4·6 0	Cantilevered from top and bottom length = 4·8 m	508 406
12·0	5	6·0 max. 4·6	4·6 0	Cantilevered from top length = 7·2 m Cantilevered from bottom length = 2·4 m	813 711

Height of assembly m	Number of tiers	Area of individual plates m²		Stabiliser arrangement	Width of stabiliser required mm
Wind load 1000 N/m²					
5·0	2 min.	6·7 max.	0	Cantilevered from top length = 2·5 m	406
7·5	3 min.	6·7 max.	0	Cantilevered from top and bottom length = 2·5 m	406
7·5	3 min.	6·7 max. 5·2 3·7	5·2 3·7 0	Cantilevered from top length = 5·0 m	610 508 406
7·5	4	5·0 max. 4·5	4·5 0	Cantilevered from top length = 3·75 m Cantilevered from bottom length = 1·9 m	508 406
9·0	4 min.	6·2 max. 4·2	4·2 0	Cantilevered from top length = 6·75 m	711 610
9·0	4 min.	6·2 max. 5·6	5·6 3·7	Cantilevered from top length = 4·5 m	610 508
		3·7	0	Cantilevered from bottom length = 2·25 m	406
12·0	5 min.	5·9 max. 5·2	5·2 3·7	Cantilevered from top length = 7·2 m	813 711
		3·7	0	Cantilevered from bottom length = 2·4 m	610
12·0	5 min.	5·9 max. 5·2	5·2 0	Cantilevered from top and bottom length = 4·8 m	610 508

Height of assembly m	Number of tiers	Area of individual plates m²		Stabiliser arrangement	Width of stabiliser required mm
Wind load 1200 N/m²					
5·0	2 min.	5·9 max.	0	Cantilevered from top length = 2·5 m	406
7·5	3 min.	5·9 max.	0	Cantilevered from top and bottom length = 2·5 m	406
7·5	3 min.	5·9 max. 4·5	4·5 0	Cantilevered from top length = 5·0 m	610 508
7·5	4	4·5 max. 3·3	3·3 0	Cantilevered from top length = 3·75 m Cantilevered from bottom length = 1·9 m	508 406
9·0	4 min.	5·6 max. 4·8 3·4	4·8 3·4 0	Cantilevered from top length = 6·75 m	813 711 610
9·0	4 min.	5·6 max. 4·2	4·2 0	Cantilevered from top length = 4·5 m Cantilevered from bottom length = 2·25 m	610 508
12·0	5 min.	5·9 max. 4·5	4·5 0	Cantilevered from top and bottom length = 4·8 m	610 508
12·0	5 min.	5·9 max. 4·5	4·5 0	Cantilevered from top length = 7·2 m Cantilevered from bottom length = 2·4 m	813 711

Propped cantilevered stabiliser system

Height of assembly m	Number of tiers	Area of individual plates m²		Width of stabiliser required mm
Wind load 700 N/m²				
5·0	2 min.	7·4 max.	0	406
7·5	3 min.	7·4 max.	0	406
7·5	4	5·6 max.	0	406
9·0	3 min.	7·4 max.	0	406
9·0	4	7·0 max.	0	406
12·0	4 min.	7·4 max.	4·6	508
		4·6	0	406
12·0	5	5·9 max.	0	508
Wind load 1000 N/m²				
5·0	2 min.	7·4 max.	0	406
7·5	3 min.	7·4 max.	0	406
7·5	4	5·6 max.	0	406
9·0	3	7·4 max.	5·6	508
		5·6	0	406
9·0	4	7·0 max.	4·2	508
		4·2	0	406
12·0	4 min.	7·4 max.	6·5	610
		6·5	0	508
12·0	5	5·9 max.	4·5	610
		4·5	0	508
Wind load 1200 N/m²				
5·0	2 min.	6·9 max.	0	406
7·5	3 min.	6·9 max.	5·9	508
		5·9	0	406
7·5	4	5·1 max.	4·5	508
		4·5	0	406
9·0	3 min.	7·4 max.	4·6	508
		4·6	0	406
9·0	4	6·4 max.	3·4	508
		3·4	0	406
12·0	4 min.	7·4 max.	4·6	610
		4·6	0	508
12·0	5	5·9 max.	5·2	711
		5·2	3·7	610
		3·7	0	508

Height of assembly m	Number of tiers	Area of individual plates m²		Width of stabiliser required mm
Wind load 1400 N/m²				
5·0	2 min.	6·4 max.	0	406
7·5	3 min.	6·4 max.	4·5	508
		4·5	0	406
7·5	4	4·8 max.	3·3	508
		3·3	0	406
9·0	3 min.	7·4 max.	6·5	610
		6·5	0	508
9·0	4	5·9 max.	4·8	610
		4·8	0	508
12·0	4 min.	7·4 max.	6·5	711
		6·5	0	610
12·0	5	6·1 max.	4·5	711
		4·5	0	610
Wind load 1700 N/m²				
5·0	2 min.	5·9 max.	0	406
7·5	3 min.	5·9 max.	3·7	508
		3·7	0	406
7·5	4	4·4 max.	0	508
9·0	3 min.	7·3 max.	5·6	610
		5·6	0	508
9·0	4	5·5 max.	4·8	610
		4·8	0	508
12·0	4 min.	7·3 max.	5·6	711
		5·6	0	610
12·0	5	5·9 max.	5·2	813
		5·2	3·7	711
		3·7	0	610
Wind load 1900 N/m²				
5·0	2 min.	5·5 max.	0	406
7·5	3 min.	5·5 max.	3·7	508
		3·7	0	406
7·5	4	4·1 max.	0	508
9·0	3 min.	6·9 max.	4·6	610
		4·6	0	508
9·0	4	5·1 max.	0	610
12·0	4 min.	6·9 max.	4·6	711
		4·6	0	610
12·0	5	5·5 max.	4·5	813
		4·5	0	711

Simply supported/pin jointed stabiliser system

Height of assembly m	Number of tiers	Area of individual plates m²		Width of stabiliser required mm
Wind load 700 N/m²				
5·0	2 min.	7·4 max.	0	406
7·5	3 min.	7·4 max.	6·7	508
		6·7	0	406
7·5	4	5·6 max.	4·5	508
		4·5	0	406
9·0	3 min.	7·4 max.	4·6	508
		4·6	0	406
9·0	4	7·0 max.	5·6	610
		5·6	0	508
12·0	4 min.	7·4 max.	6·5	711
		6·5	0	610
12·0	5	6·0 max.	5·2	711
		5·2	0	610
Wind load 1000 N/m²				
5·0	2 min.	7·4 max.	0	406
7·5	3 min.	7·4 max.	4·5	508
		4·5	0	406
7·5	4	5·6 max.	3·3	508
		3·3	0	406
9·0	3 min.	7·4 max.	6·5	610
		6·5	0	508
9·0	4	7·0 max.	6·3	711
		6·3	4·2	610
		4·2	0	508
12·0	4 min.	7·4 max.	6·5	813
		6·5	5·6	711
		5·6	0	610
12·0	5	6·0 max.	5·2	813
		5·2	3·7	711
		3·7	0	610
Wind load 1200 N/m²				
5·0	2 min.	7·0 max.	0	406
7·5	3 min.	7·0 max.	6·7	610
		6·7	3·7	508
		3·7	0	406
7·5	4	5·2 max.	4·5	610
		4·5	0	508
9·0	3 min.	7·4 max.	4·6	610
		4·6	0	508
9·0	4	6·5 max.	4·9	711
		4·9	0	610
12·0	4 min.	7·4 max.	5·6	813
		5·6	0	711
12·0	5	6·0 max.	4·5	813
		4·5	0	711
Wind load 1400 N/m²				
5·0	2 min.	6·3 max.	0	406
7·5	3 min.	6·3 max.	5·2	610
		5·2	0	508
7·5	4	4·7 max.	3·9	610
		3·9	0	508
9·0	3 min.	7·4 max.	6·5	711
		6·5	0	610
9·0	4	5·9 max.	4·2	711
		4·2	0	610
12·0	4 min.	7·4 max.	5·6	914
		5·6	4·6	813
		4·6	0	711
12·0	5	6·0 max.	5·2	914
		5·2	3·7	813
		3·7	0	711
Wind load 1700 N/m²				
5·0	2 min.	5·9 max.	5·2	508
		5·2	0	406
7·5	3 min.	5·9 max.	4·5	610
		4·5	0	508
7·5	4	4·4 max.	0	610
9·0	3 min.	7·3 max.	5·6	711
		5·6	0	610
9·0	4	5·5 max.	4·8	813
		4·8	0	711
12·0	4 min.	7·3 max.	6·5	1016
		6·5	4·6	914
		4·6	0	813
12·0	5	5·9 max.	5·2	1016
		5·2	4·5	914
		4·5	0	813
Wind load 1900 N/m²				
5·0	2 min.	5·5 max.	4·5	508
		4·5	0	406
7·5	3 min.	5·5 max.	3·7	610
		3·7	0	508
7·5	4	4·0 max.	0	610
9·0	3 min.	6·9 max.	4·6	711
		4·6	0	610
9·0	4	5·1 max.	4·2	813
		4·2	0	711
12·0	4 min.	6·9 max.	5·6	1016
		5·6	4·6	914
		4·6	0	813
12·0	5	5·5 max.	4·5	1016
		4·4	3·7	914
		3·7	0	813

Glass specification

Definitions

Whereas annealed glass is virtually stress free, toughened glass has stable balanced stresses induced deliberately by a heating and cooling treatment which results in high compression at the surfaces, with compensating tension in the centre giving an increased resistance to mechanical and thermal stressing. The toughening process can be applied to clear glass, to surface-modified Float glass, or to body-tinted heat-absorbing glass.

Maximum bow in toughened glass

Side length mm	Nominal thickness 10 mm	12 mm and over
up to 900	0– 3 mm	0– 3mm
901–1200	0– 4 mm	0– 4 mm
1201–1500	0– 6 mm	0– 6 mm
1501–1800	0– 9 mm	0– 9 mm
1801–2100	0–11 mm	0–10 mm
2101–2400	0–14 mm	0 12 mm
2401–2700	0–16 mm	0–14 mm
2701–3000	0–18 mm	0–16 mm
3001–3300	0–21 mm	0–18 mm
3301–3600	0–24 mm	0–21 mm
3601–3950	0–28 mm	0–25 mm

Thickness tolerance

Nominal thickness mm	Tolerance mm
10	9·7–10·3
12	11·7–12·3
15	14·7–15·3
19	18·2–19·9

Dimensional tolerances

Nominal thickness mm	Size mm	Tolerance mm
10	up to 1200	+1 −2
	over 1200	+2 −2
12	up to 1200	+1 −2
	over 1200	+2 −2
15	up to 1200	+1 −2
	over 1200	+2 −2
19	up to 1200	+1 −2
	over 1200	+2 −2

General information

Handling. As with all glass, the edge of toughened glass is vulnerable and care is necessary both in handling and installation, since an edge damaged in handling may result in subsequent breakage.

Storage. Toughened glass should be stored in a near vertical position, set on edge on strips of wood, felt or other relatively soft material. It should not be stacked horizontally.

Tongs marks. These are small indentations which are introduced during processing. In some cases they are not present. Tongs marks are normally kept as near to the edge as possible, their position being dependent on shape, thickness and other factors.

Where the major dimension of the glass exceeds 2500 mm the tongs marks will be on the long edge. Where the major dimension does not exceed 2500 mm the tongs marks may be on either edge. Where there are specific requirements advice should be sought.

Maximum available sizes of toughened glass

Glass	Nominal thickness mm	Length mm	Width mm	Notes
Clear	10	3950*	1520	
		3100*	2410	
	15	3950*	1520	max 5·0 m²
		3100*	2410	max 6·7 m²
	19	3950*	1520	max 4·2 m²
		3100*	2410	max 5·6 m²
Surface-modified Float	10	2700	2000	
	12	2700	2000	
Body-tinted heat-absorbing grey or bronze	10	2700	2000	
	12	3100	2410	

*Where one dimension exceeds 1520 mm the other cannot exceed 3100 mm.

Properties

Strength. The degree to which the resistance to applied loading of the glass is increased by toughening can be varied but is dependent to some extent on the thickness and nature of the glass. In general, the strength can be increased by up to four to five times. The toughening process does not increase the resistance of the surface to scratching or abrasion.

Thermal resistance. Toughened glass offers greatly increased resistance to sudden temperature changes compared with annealed glass. The stress characteristics of toughened glass of normal soda-lime composition are unchanged up to about 300°C and are not affected by sub-zero temperatures. It can be exposed to a thermal gradient so that one surface of the glass reaches a temperature not exceeding 250°C, while the other is exposed to ambient air temperatures.

Fracture. When broken, toughened glass fractures into small particles which are unlikely to inflict cuts as caused by broken annealed glass.

Working of toughened glass. Toughened glass must not be cut or worked. This will affect its properties and may result in breakage. Therefore, all work must be done before toughening.

Specification

Flatness. By nature of the toughening process it is not possible to produce consistently glass as flat as annealed glass. The deviation depends on thickness, size, aspect ratio and other factors. The table below specifies the maximum bow for any particular size and thickness. Bow is measured with the glass in the vertical plane and supported at quarter points. The maximum deviation from a straight edge is measured on the concave surface. For glasses with aspect ratios between 1·1 and 1·3:1 tolerances should be increased by 1 mm.

'Tongs kink'. Due to processing requirements some local deviations may occur on one edge of a plate. This deviation, or 'tongs kink', may be up to 2 mm in any 300 mm length measured 25 mm in from the edge of the glass.

Appendix II

Glossary of technical terms

Absorptance
The fraction of incident radiation that is absorbed by a surface or in a layer of glass.

Aspect ratio
The ratio of the longer to the shorter side of a pane of glass.

Back clearance
The clearance between a pane of glass and the upstand of the rebate.

Basic wind speed
The speed that is appropriate to the wind in a given region: it is modified by local factors for design purposes.

British Zonal classification
The number, eg BZ5, that classifies the downward distribution of light from a source in the British Zonal method.

Celsius
In the International System of Units (SI), temperatures are expressed in degrees Celsius (°C). Differences of temperature are expressed in kelvins (K). The two units have the same size, so that the difference between an indoor temperature of 20 degrees Celsius and an outdoor temperature of 5 degrees Celsius is 15 kelvins: 20°C−5°C=15K.

CIE sky
An overcast sky for which the variation of luminance with altitude is that adopted as standard by the Commission Internationale de l'Eclairage. The sky luminance, L_a, at an altitude, a, is given by $L_a = L_Z(1+2 \sin a)/3$ where L_Z is the sky luminance at the zenith.

Cylindrical illuminance
See illuminance.

Daylight factor
The ratio (expressed as a percentage) of the daylight illuminance at a point to the illuminance received simultaneously on a horizontal plane exposed to a complete, unobstructed sky, direct sunlight being excluded from both measurements.

Decibel
The unit of sound level. It is a logarithmic unit, based on ratios, that approximately describes the response of the ear. A sound intensity, I_1, is greater than a sound intensity, I_2, by LdB when $10 \log_{10} I_1/I_2 = \text{LdB}$.

Design wind speed
Derived from the basic wind speed by applying factors to allow for local topology, ground roughness and the expected life of the building.

Direct transmittance
See transmittance.

Distance piece
A resilient block used to determine the back clearance or the face clearance.

Edge clearance
The clearance between the edge of the glass and the platform of the rebate: usually, half the difference between the tight size and the glazing size.

Environmental temperature
The temperature of the hypothetical uniform environment (with surroundings and air at equal temperatures) that would give the same rate of heat transfer through a building element as occurs under the prevailing conditions. It is the sum of two-thirds of the mean radiant temperature and one-third of the air temperature.

Externally reflected component
The part of the daylight factor that is due to light reaching the reference point after reflection from an outdoor surface, but with no indoor reflection.

Face clearance
The clearance between a pane of glass and a retaining bead.

Flux fraction ratio
The ratio of the luminous flux from a source emitted into the upper hemisphere to that emitted into the lower hemisphere.

Glazing size
The dimension of a pane of glass. It is less than the tight size by the sum of the edge clearances.

Gnomon
The pin of a sun dial. The direction and length of its shadow indicate the azimuth and altitude of the sun.

Gnomonic projection
A projection based on the same principles as a sun dial. Simple perspective.

Hemispherical illuminance
See illuminance.

Horizontal illuminance
See illuminance.

Illuminance
The amount of luminous flux falling on unit area. It is expressed in lux.

Illuminance, horizontal
The illuminance on a horizontal plane at the reference point.

Illuminance, mean cylindrical
The average illuminance on the curved surface of an infinitesimal cylinder with its axis vertical at the reference point.

Illuminance, mean hemispherical
The average illuminance on the surface of an infinitesimal hemisphere at the reference point. The response of a photoelectric cell that measures the mean hemispherical illuminance, E_d, due to a small source that produces an illuminance E_n measured normal to the direction of the source is given by: $E_d = 0{\cdot}25 \,.\, E_n(1+\cos\theta)$, where θ is the angle between the axis of the cell and the direction of the source.

Illuminance, mean spherical
The average illuminance on the surface of an infinitesimal sphere at the reference point.

Illuminance, scalar
The average value of the illumination solid, equivalent to mean spherical illuminance.

Illumination
The physical process of lighting. The term no longer applies to the measurement of luminous flux reaching a surface (*see* illuminance).

Illumination solid
A three-dimensional solid for which the distance from the reference point to the surface of the solid in any direction is proportional to the illuminance on a surface perpendicular to that direction.

Illumination vector
The magnitude and direction of the greatest difference in illuminance in opposite directions at a point. The British Standards Institution recommends that bold italic symbols should be used for vector quantities and, although $\boldsymbol{E}$ is readily distinguished from E when they appear in print, to identify the vector quantity in manuscript a small arrow may be used over the letter thus $\vec{E}$.

Internally reflected component
The part of the daylight factor that is due to light reaching the reference point after reflection from one or more of the internal surfaces.

Kelvin
The SI unit of temperature difference (*see* Celsius).

Light transfer ratio
A function of the geometry of a roof-light system that evaluates the transmitted light in relation to the incident light.

Location block
A resilient block used to determine the edge clearance along the top or vertical sides of a pane of glass. The same function along the bottom is performed by setting blocks.

Long wave shading coefficient
See shading coefficient.

Lumen
The fundamental unit measuring the rate of flow of light by its visual effect, independently of the colour of the light. One lumen is the light flux emitted in unit solid angle by a point source of unit intensity.

Luminance
Expresses the amount of light emitted in a given direction by a source or a reflecting surface in relation to its apparent size. The unit is the candela per square metre.

Lux
The unit of illuminance; one lumen per square metre.

Mean radiant temperature
The temperature of a uniform black enclosure in which a solid body or occupant would exchange the same amount of radiant heat as in the existing non-uniform environment. It is closely related to the mean temperature of the surfaces.

Mean sound insulation
The average of the sound reduction index over the frequency range from 100 to 3150 hertz.

Noise climate
The noise level that is exceeded for only 10 per cent of the period from 0600 to 2400 hours on a normal working day.

Overcast sky
See CIE sky.

Partition factor
The fraction of the solar radiant energy absorbed by the glass that is admitted to the building.

P.s.a.l.i.
Permanent supplementary artificial lighting of interiors.

Rebate
The recess in a window frame, fashioned to receive the edge of the glass.

Rebate depth
The dimension of the upstand of the rebate, parallel to the plane of the glass.

Rebate width
The dimension of the rebate platform, perpendicular to the plane of the glass.

Reflectance
The fraction of incident radiation that is reflected from a surface.

Relative humidity
The ambient water vapour pressure expressed as a percentage of the saturation water vapour pressure at the same temperature.

Room index
The number that describes the proportions of a rectangular room

$$\text{Room index} = \frac{\text{length} \times \text{width}}{(\text{length} + \text{width}) \times \text{height above working plane}}$$

Scalar illuminance
See illuminance.

Shading coefficient
A number used to compare the solar radiant heat admission properties of different glazing systems. It is calculated by dividing the appropriate transmittance by 0·87 which is the total transmittance of a notional clear single glazing between 3 and 4 mm thick.

Shading coefficient, long wave
The fraction of the absorptance that contributes to the total transmittance divided by 0·87.

Shading coefficient, short wave
The direct transmittance divided by 0·87.

Shading coefficient, total
The total transmittance divided by 0·87.

Sight size
The dimension of the opening that admits light. It is less than the tight size by the sum of the rebate depths.

Sky component
The part of the daylight factor that is due to the light reaching the reference point directly from the sky without reflection from any external or internal surface.

Solar radiant heat
Radiation having the spectral distribution defined by Moon's curve for air mass 2. This is approximately the same as the total radiation (ultra-violet, visible, and infra-red) that is received at sea level directly from the sun at an altitude of 30°.

Spherical illuminance
See illuminance.

Sun path
The apparent track of the sun across the sky during a given day.

Thermal conductivity
The rate of transfer of heat through unit area of a material between surfaces that are unit distance apart and have unit difference of temperature. SI unit=W/mK.

Thermal transmittance
The rate of transfer of heat through unit area of a structure for unit difference of air temperature on the two sides. SI unit= W/m^2K. Commonly called U-value.

Tight size
The dimension of the window frame between the platforms of the rebates. It must be reduced by the sum of the edge clearances to obtain the glazing size.

Total transmittance
See transmittance.

Transmittance
The fraction of the incident radiation that is transmitted through the glass.

Transmittance, direct
The fraction of solar radiant heat at normal incidence that is transmitted directly through the glazing without change of wavelength.

Transmittance, thermal
See thermal transmittance.

Transmittance, total
The fraction of solar radiant heat at normal incidence that is transferred through the glazing by all means. It is composed of the direct transmittance and an appropriate fraction of the absorptance.

Transmittance, visible
The fraction of visible light at normal incidence that is transmitted through the glazing.

Utilisation factor
The portion of the total luminous flux that reaches the working plane.

U-value
See thermal transmittance.

Vector/scalar ratio
The ratio of the illumination vector to the scalar illuminance. It indicates the degree of asymmetry of the illumination solid and is thus the index of the strength of the flow of light.

Vector solid
A sphere that represents the asymmetrical component of the illumination solid. Its surface contains the reference point.

Visible light
Light having a spectral distribution corresponding to the CIE (Commission Internationale de l'Eclairage) Standard Illuminant C. This is approximately the same as daylight.

Visible transmittance
See transmittance.

Wind speed, basic
See basic wind speed.

Wind speed, design
See design wind speed.

Working plane
A horizontal plane for reference purposes: assumed to be 0·85 m above floor level when no better information is available.

Appendix III

Bibliography

Environmental design

1 *Office design: a study of environment* P. Manning (ed). Pilkington Research Unit, Department of Building Science, University of Liverpool. 1965.
2 *Heat reflecting/low light transmission glasses in office building* P. G. T. Owens (ed). *The Architects' Journal*, 8.5.74.
3 *The significance of sunshine and view for office workers* CIE Intersessional Conference on Sunlight in Buildings, Newcastle upon Tyne, 1965. Proceedings published by Bouwcentrum, Rotterdam, 1967.
4 Keighley, E. C. *Visual requirements and reduced fenestration in offices* Building Research Establishment Current Paper 41/74.
5 *Sunlight and daylight* DOE Welsh Office. HMSO 1971.
6 The Illuminating Engineering Society. *Daytime lighting in buildings* IES Technical Report No. 4. Second edition. London, 1972.
7 British Standard Code of Practice, CP3. *Code of basic data for the design of buildings* (Various separate chapters).
8 *Windows and environment* Pilkington Brothers Limited. 1969.
9 Cuttle, C. *The use of special performance glazing materials in modern offices* MA Thesis, University of Manchester. 1974.
10 The Institution of Heating and Ventilating Engineers. *IHVE Guide, Book A* London 1970 and 1975.
11 Jennings, R. and Wilberforce, R. R. 'Thermal comfort and space utilisation'. *Insulation* March, 1973, pp 57-60.
12 Lynes, J. A., Burt, W., Jackson, G. K. and Cuttle, C. *The flow of light into buildings. Transactions of the Illuminating Engineering Society (London)*. Vol 31, no. 3, 1966.
13 The Illuminating Engineering Society. *The IES Code for interior lighting* London. 1977
14 Givoni, B. *Man, climate and architecture* Elsevier Architectural Science Series. 1969.
15 The American Society of Heating, Refrigerating and Air-Conditioning Engineers. *Handbook of fundamentals* 1972.
16 *The primary school: an environment for education* P. Manning (ed). Pilkington Research Unit. Department of Building Science, University of Liverpool, 1967.

Lighting design

1 *Sunlight and daylight* DOE Welsh Office. London HMSO, 1971.
2 *The flow of light in lighting design* Environmental Advisory Service Report. Pilkington Brothers Limited, 1974.
3 The Illuminating Engineering Society. *Daytime lighting in buildings* IES Technical Report No. 4. Second edition. London, 1972.
4 British Standard Code of Practice, CP3. *Code of basic data for the design of buildings* (Separate chapters on various aspects of lighting design).
5 Cuttle, C. and Slater, A. I. 'A low-energy approach to office lighting'. *Light and Lighting* January/February 1975.
6 *Estimating daylight in buildings—*1 Building Research Station Digest 41. London HMSO 1970.
7 Hopkinson, R. G. *Architectural physics: lighting* London HMSO. 1963.
8 The Illuminating Engineering Society. *The IES Code for interior lighting* London. 1977.
9 *Colour discrimination and heat rejecting window glasses* Internal Report No. 10/73. Plymouth Polytechnic School of Architecture.
10 Lynes, J. A., Burt, W., Jackson, G. K. and Cuttle, C. 'The flow of light into buildings *Transactions of the Illuminating Engineering Society (London)*. Vol. 31, no. 3, 1966.
11 Cuttle, C., Valentine, W. B., Lynes, J. A. and Burt, W. *Beyond the working plane* Paper P-67.12 in the Proceedings of the CIE Conference, Washington, 1967.
12 Cuttle, C. 'Lighting patterns and the flow of light'. *Lighting Research and Technology* Vol. 3, no. 3, 1971.
13 *The calculation of utilization factors—the BZ method* London. The Illuminating Engineering Society. February 1971. IES Technical Report No. 2.
14 Hopkinson, R. G. and Longmore, J. 'The permanent supplementary artificial lighting of interiors' *Transactions of the Illuminating Engineering Society (London)*. 1959.
15 Owens, P. G. T. *Energy conservation and office lighting* CIB symposium on energy conservation in the built environment. Watford, 1976.
16 *Windows and environment* Pilkington Brothers Limited. 1969.
17 Walsh, J. W. T. *The science of daylight* London, 1961.
18 Manning, P. *The design of roofs for single-storey general-purpose factories* Liverpool, The Department of Building Science, University of Liverpool, 1962.
19 Hopkinson, R. G. *Hospital lighting* London, 1964.

Thermal design

1 Petherbridge, P. *Sunpath diagrams and overlays for solar heat gain calculations* Building Research Establishment Current Papers, Research Series 39. 1965.
2 *Solar heat gain through windows* Environmental Advisory Service Report. Pilkington Brothers Limited. Fourth edition, 1974.
3 The Institution of Heating and Ventilating Engineers. *IHVE Guide* London. 1970 and 1975.
4 *Solar control performance of blinds* Environmental Advisory Service Report. Pilkington Brothers Limited. July, 1973.
5 Owens, P. G. T. and Barnett, M. 'Reducing glazing U-values with low emissivity coatings and low conductivity gases'. *Building Services Engineer* February and March 1974.
6 *The building (second amendment) regulations* 1974 Statutory Instruments No. 1944. HMSO.
7 Jennings, R. and Wilberforce, R. R. 'Thermal comfort and space utilisation'. *Insulation* March 1973.
8 *Air-conditioning loads by computer* Environmental Advisory Service Report. Second edition. Pilkington Brothers Limited. April, 1974.

9 *Obstructional shading by computer* Environmental Advisory Service Report. Pilkington Brothers Limited. May. 1973.
10 Wilberforce, R. R. *The effect of solar radiation on window energy balance* International CIB Symposium, Energy Conservation in the Built Environment. April, 1976.
11 Milbank, N. O. and Harrington-Lynn, J. 'Thermal response and the admittance procedure'. *Building Services Engineer* May, 1974.
12 *Thermal transmission of windows* Environmental Advisory Service Report. Pilkington Brothers Limited. Third edition, March, 1973.
13 *Tables of temperature, relative humidity, precipitation and sunshine for the world* Meteorological Office. Met.O. 856. London. HMSO.
14 Wilberforce, R. R. 'The energy balance of glazing'. *Building Services Engineer* March, 1976.
15 Moon, P. 'Proposed standard radiation curves for engineering use'. *Journal of the Franklin Institute* Vol. 230. November, 1940.
16 The American Society of Heating, Refrigerating and Air Conditioning Engineers. *Handbook of fundamentals* 1972.
17 *Windows and environment* Pilkington Brothers Limited. 1969.
18 Langdon, F. J. *Modern offices: a user survey* London, HMSO, 1966. National Building Studies. Research paper no. 41.
19 British Standard Code of Practice, CP3. *Code of basic data for the design of buildings* (Various separate chapters on thermal topics).
20 *Homes for today and tomorrow* Ministry of Housing and Local Government. London, HMSO, 1961.

Acoustic design

1 British Standards, BS 3489. *Specification for sound level meters (industrial grade)* British Standards Institution, 1962.
2 British Standards, BS 2475. *Specification for octave and one-third octave band-pass filters* British Standards Institution, 1964.
3 Kosten, C. W. and Van Os, G. J. *Community reaction criteria for external noises* NPL Symposium No. 12, The Control of Noise. London, HMSO, 1962.
4 Parkin, P. H., Purkis, H. J., Stephenson, R. J. and Schlaffenberg, B. *London noise survey* London, HMSO, 1968.
5 *Noise and buildings* Building Research Station Digest (Second Series) No. 38, 1963.
6 *Windows and environment* Pilkington Brothers Limited. 1969.
7 *Glass and noise control* Pilkington Brothers Limited. May 1976.
8 Parkin, P. H. and Humphreys, H. R. *Acoustics noise and buildings* Third edition London, 1969.
9 Beranek, L. L. *Noise reduction* New York, 1960.
10 *Noise. Final report of the committee on the problem of noise* Cmnd 2056. London, HMSO, 1963.
11 Harris, C. M. *Handbook of noise control* New York 1957.
12 British Standard Code of Practice, CP3. *Code of basic data for the design of buildings. Chapter III, Sound insulation and noise reduction* London, 1960.

Energy conservation

1 Petherbridge, P. *Sunpath diagrams and overlays for heat gain calculations* Building Research Establishment. Current Papers, Research Series 39, 1965.
2 Cuttle, C. and Slater, A. I. 'A low energy approach to office lighting'. *Light and Lighting*. January/February 1975.
3 Burberry, P. 'Conserving energy in buildings'. *The Architects' Journal* 11.9.74.
4 *The building (second amendment) regulations* 1974 Statutory Instruments No. 1944, HMSO.
5 Wilberforce, R. R. *The effect of solar radiation on window energy balance* International CIB Symposium, Energy Conservation in the Built Environment. April, 1976.
6 Wilberforce, R. R. 'The energy balance of glazing'. *Building Services Engineer* March, 1976.
7 Milbank, N. O., Dowdall, J. P. and Slater, A. *Investigation of maintenance and energy costs for services in office buildings* Building Research Station Current Paper 38, 1971.
8 *Energy conservation: a study of energy consumption in buildings and possible means of saving energy in housing* Building Research Station Current Paper 56, 1975.
9 Owens, P. G. T. *Energy conservation and office lighting* CIB Symposium on Energy Conservation in the Built Environment. Watford, 1976.
10 Winch, G. R. and Burt, W. *Energy conservation measures in buildings—environmental design strategy* CIB Symposium on Energy Conservation in the Built Environment. Watford, 1976.

Special glasses

1 'Heat reflecting/low light transmission glasses in office building'. P. G. T. Owens (ed). *The Architects' Journal* 8.5.74.
2 *Colour discrimination and heat rejecting window glasses* Internal Report No. 10/73. Plymouth Polytechnic School of Architecture.
3 Cuttle, C. *The use of special performance glazing materials in modern offices* MA Thesis. University of Manchester. 1974.
4 Owens, P. G. T. and Barnett, M. 'Reducing glazing U-values with low emissivity coatings and low conductivity gases'. *Building Services Engineer* February and March, 1974.
5 *Thermal transmission of windows* Environmental Advisory Service Report. Pilkington Brothers Limited. Third edition. March, 1973.
6 *The application of solar control glasses* Pilkington Brothers Limited. January, 1972.
7 *Glass and insulation* Pilkington Brothers Limited. May, 1975.

Structural design

1 *Glazing manual* The Flat Glass Association. London, 1968. (Second edition in preparation for publication by the Glass and Glazing Federation).
2 British Standard Code of Practice, CP 152: 1972. *Glazing and fixing of glass for building*.
3 British Standard Code of Practice, CP3. *Code of basic data for the design of buildings* (Various separate chapters and parts on loading).
4 British Standard BS 952: 1964. *Classification of glass for glazing and terminology for work on glass*.
5 Fairweather, L. and Sliwa, Jan A. *AJ Metric Handbook* London, The Architectural Press. Second revised edition 1969.
6 British Standard BS 4255: Part 1: 1967. *Preformed rubber gaskets for weather exclusion from buildings*.
7 *Potential for thermal breakage of sealed double glazing units* Canadian Building Digest 129. Ottawa, National Research Council of Canada, September 1970.
8 McGrath, R. *Glass in architecture and decoration* London. The Architectural Press, 1961.
9 Beckett, H. E. and Godfrey, J. A. *Windows—performance, design and installation* Crosby Lockwood Ltd. 1974.

Index